Reconstituting rurality

Class, community and power in the development process

JONATHAN MURDOCH
Centre for Rural Economy,
University of Newcastle upon Tyne

TERRY MARSDEN
School of Geography and Earth Resources,
University of Hull

Routledge
Taylor & Francis Group

LONDON AND NEW YORK

First published in 1994 by UCL Press

Reprinted 2003 by Routledge
11 New Fetter Lane
London, EC4P 4EE

*Routledge is an imprint of the
Taylor & Francis Group*

British Library Cataloguing in Publication Data
A catalogue record for this book is available from the British Library.

Library of Congress Cataloging-in-Publication data are available.

ISBN: 1-85728-645-6 PB

Printed and bound by Antony Rowe Ltd, Eastbourne

Contents

Preface

This is the second book in the Restructuring Rural Areas Series from the London Countryside Change Programme, sponsored by the Economic and Social Research Council (ESRC). In the first volume, *Constructing the countryside*, we, along with our colleagues, proposed a new set of theoretical and methodological perspectives on rural change. Although that volume sketched in the historical background to present changes, it included no new empirical work. In this book we present the first of our locality studies where we take forward our theoretical and methodological perspectives using material collected during the period 1989–92 in the district of Aylesbury Vale in Buckinghamshire. We provide here a set of detailed studies of the land-development process in this rural locality in order to illustrate the most salient contemporary forces of change in this and other similar rural areas.

In conducting this study we have been helped by many people in both the locality and in our places of work. We wish to thank particularly our many respondents in Aylesbury Vale who gave their time to talk to us and who are too numerous to mention here. We are particularly grateful to staff in Buckinghamshire County Council and Aylesbury Vale District Council Planning Departments. We would simply not have been able to produce this book without their help. We are also grateful to our colleagues on the Countryside Change Programme – Andrew Flynn, Julie Grove-Hills, Philip Lowe, and Richard Munton – who worked closely with us during the time of this study. Further thanks are also due to Graham Day, Liz Hawkins, Andy Pratt, Neil Ward and Sarah Williams. We are also grateful to the ESRC for funding this work.

Introduction

The study of structural social inequality ... remains a central feature of the sociological enterprise.   *Rosemary Crompton* (1993)

Being comfortable, many raise vigorous objection to that which invades comfort. What is important is that there is no self-doubt in their present situation. The future for the contented majority is thought effectively within their personal command. Their anger is evident – and, indeed, can be strongly evident – only when there is a threat or possible threat to present wellbeing and future prospect – when government and the seemingly less deserving intrude or threaten to intrude their needs and demands ...

However intervention by the state may be condemned in the age of contentment, it has been relatively comprehensive when the interests of the contented are involved and relatively limited when the problems are those of the poor."   *J. K. Galbraith* (1992)

Good God, is a country like this to be ruined by the folly of those who govern it?   *William Cobbet* on Aylesbury Vale (1829)

We begin with two anecdotes. Driving into a picturesque, self-contained, affluent village we were immediately struck by how *visible* we were. The few people on the street stopped to regard us with quiet intent; we were, of course, strangers. We were in the village to interview the local vicar and this was our first experience of fieldwork in this particular place. Having located the vicarage, nicely situated opposite the village pond and the manor house, we pulled into the driveway, got out of the car and rang the doorbell. There was no reply, so we waited for a couple of minutes and rang the doorbell again, but at that moment a large estate car pulled very sharply into the drive and a middle-aged woman jumped out, demanding to know what we wanted. The vicar apparently no longer lived in the vicarage, he lived in a new bungalow further down the street. The woman brusquely told us how to find his house and waited for us to leave. By this time our feeling of visibility had become extremely acute. Clearly, this woman had not returned home at that point coincidentally; she had known there were two strangers standing at her front door. Quite how she had been alerted so quickly remained a mystery to us.

The second anecdote refers to an experience in another village. It was a year or so later. Having completed the bulk of our interviews we decided to take advantage of a clear summer's day to take some photographs of the village. We parked our car in the centre and casually wandered around, photographing key development sites. After about ten minutes we became aware of a woman running down the street towards us, shouting. When she reached us she demanded to know what we were doing. She was the local schoolteacher and had told the schoolchildren to watch out for strangers in the village. They had reported the fact that there were two strange men taking photographs, and she was spurred into action and we were asked to explain ourselves. Again, we felt like intruders.

We recount these two incidents because they touch on many of the concerns of this book. Villages are small scale and partially enclosed "communities". Entering a village one becomes aware of how "public" spaces can also be "private". As a stranger you may enter but only on the understanding that you do not loiter and that you have a purpose. The two experiences recounted here illustrate how quickly these purposes may be demanded of the "stranger" and this, in turn, reflects the prevailing ethos of community life in the village and the way in which village membership is asserted. In both of the incidents described, our interrogators were by most conventional criteria (well spoken, smartly dressed, self-confident) "middle class", and the tone in which it was demanded of us that we explain ourselves was that of self-righteousness, bordering on indignation. An adequate response was expected and we were left in no doubt that we were disturbing the "natural rhythm" of the place.

These villages are in many senses beautiful places; they have ponds, trees surround many of the houses, traditional buildings and areas of green space constitute the central aspects, and an air of tranquillity is imposed to maintain the facade of timeless arcadian village life, so craved by those who have in recent times actively sought a "place in the country". Indeed, we might speculate that the extent to which villages maintain their traditional material shape and appearance is in inverse proportion to the amount of social change that has taken place.

Of course, these experiences were to be expected. It could be argued that we, as researchers, provoked them, or at least sought them out. We were conducting fieldwork in the district of Aylesbury Vale in Buckinghamshire, approximately forty miles to the northwest of London. Villages such as these have been home to a commuting population since at least the 1930s (Lewis 1962), and provide fruitful places to examine the long-term effects of middle-class occupation in what *looks* like a traditional rural area. In fact, so associated with commuters and ex-urban middle-class residents has the outer South East become that it is sometimes doubted whether this part of England is "rural" in any meaningful sense at all. Thrift, for instance, treats it as "one vast created and manicured urban/suburban space" (1987b: 77). Leaving aside for the moment how we might define the "rural", we feel unable to subscribe to the view that this whole geographical area – which includes fields, woods, housing estates, roads, parks, retail parks, industrial estates, gardens, rivers, etc. – can be subsumed under the simple category of "suburban". Aylesbury Vale, with the exception of the town of Aylesbury, cannot be considered in the same way as the outer London suburbs or "satellite" towns such as Guildford and the new city of Milton Keynes. Yet Thrift's remark seems to imply that little is to be gained by differentiating "suburban" space. We believe, on the contrary, not only that suburban space is strongly differentiated but what is really interesting is the way in which economic, political and sociocultural processes give rise to, and sustain, such social and physical differentiation.

The patterns of differentiation that lead to spaces becoming "marked off" from one another indicate the need for a complex understanding of the various processes which somehow underlie the composition of particular localities and regions. We do not believe that the nature of change in places can simply be "read off" from some structural shift in social and spatial relations (such as post-Fordism or postmodernism) which somehow gives a new underlying "logic" to the range of divergent outcomes that can be discerned in practice. While it is evident that differentiation is being encouraged by new economic tendencies and cultural forms, it is equally true that these are situated in particular places at particular times and are subject to different degrees of resistance, modification and transformation.

This type of approach to rural space can be seen as distinct from many of the most influential forms of sociological and economic thought, mainly inherited from the nineteenth century. At that time it was widely believed, on the one hand, that the rural and the urban were discrete, while, on the other, it was considered that society and space would become increasingly homogeneous. In the rural domain such ideas were bound up with concerns associated with modernization and the inexorable spread of capitalist (which in the British context often meant "urban") relations. However, in more recent times the recognition of difference and diversity, particularly in the urban realm, and the increasingly divergent development trajectories of distinct localities, has brought into question many of the underlying assumptions of these earlier beliefs. The phenomenon of "differentiation" in the rural world has been partially responsible for undermining the previously dominant rural sociological paradigms associated with structural-functionalism, modernization theory and orthodox Marxism. These have given way to a much more diverse set of approaches, reflecting perhaps the (increased?) diversity of rural life. The old certainties, based upon a clear distinction between the urban and the rural, no longer hold, and thus the status of rural sociology itself is now uncertain. In this book we seek, in part, to contribute to a new focus for the subdiscipline through an examination of how the rural is made and reconstituted within one locality. We hope that the work presented here will allow us to begin to put some of the current uncertainties behind us.

Rural sociology developed throughout much of the twentieth century largely around a common set of suppositions associated with the "distinctiveness" of rural life. The first supposition concerned the notion of "traditional" rural society, while the second was linked to the changes in rural economy and society which occurred as rural areas became integrated into modern society. Often the rural was viewed as unchanging, while the "urban" (or the "modern" or the "national" – these terms were frequently used interchangeably) was seen as dynamic and expansionist, forever seeking to pull the rural into its orbit. Partly as a result of these conceptions, the lines of demarcation between the "rural" and the "urban" have been recognized by sociologists as increasingly blurred. The notion that the rural changed only at the beckoning of external forces also came into question. Agriculture for instance, the mainstay of the "rural economy", seemed to be undergoing some kind of technological revolution, thus continually substituting capital for labour, and traditional rural communities were becoming increasingly hard to distinguish. Moreover, once the surface of community life in rural villages was peeled away, a whole range of lifestyles and tensions could be discerned. The problems associated with the term "rural" came to be reflected in the ambivalent status of rural sociology itself (Marsden et al. 1990).

In recent years rural sociology has been dominated by two main sets of concerns. First, there is what has been termed "agrarian political economy", which has concerned itself primarily with the rôle of agriculture in food commodity systems. The main focus of this work has been the long-run tendency for the food industry to reduce the contribution of agriculture to the food-production process (see for instance Goodman et al. 1987). The other main approach has been derived from what is broadly termed the "restructuring thesis". Work in this vein has attempted to uncover the way rural resources – land and labour – are increasingly integrated into circuits of industrial accumulation. It challenges the specificity of rural localities, arguing that their economic complexion derives from past "rounds of investment" as capital seeks out profitable forms of production. The dominance of agriculture cannot be guaranteed, as rural areas may become attractive to other industrial sectors, promoting (uneven) development in the countryside, not simply because there are resources present that can be used in the production process, but because these areas now present consumption opportunities for a range of different income groups (Marsden 1992). Inextricably linked in some way to these economic changes have been the social, cultural and political components of life in rural localities. While the interplay between these has often been hard to discern, it is recognized that particular combinations of these processes consign rural areas to a range of different development trajectories, fostering new rounds of uneven development.

These broad approaches form the background to the first book in this series, *Constructing the countryside* (Marsden et al. 1993). In that volume an attempt was made to draw together the various strands of rural sociology into a form which would lend coherence to the contemporary analysis of rural change. Without repeating in great detail the arguments presented in the former book, we can say that, after reviewing the political economy and restructuring perspectives, we argued that the "rural" is best regarded as the outcome of a variety of economic, social and political processes and that these might usefully be observed from the vantage point of *land development*.

The essential feature of development is (usually) a change in land use, one which alters the economic, political and sociocultural relations surrounding particular pieces of land. Land use itself can be seen as a reflection of these relations. The discrete social demands on land, the political rules that surround its transfer between uses, and the tendency for capital to become "fixed" in land, have produced a series of segmented land development markets orientated towards different sectors of production and consumption. The key *rural* land-development processes are constituted within the following markets: agriculture, forestry, industry, mining, housing and leisure. The relationships between these sectors are constantly changing. For instance, in the current period, the interests of agriculture no longer occupy an unquestioned leading position as they have done over much of the countryside during the post-war period. There are growing and more widespread pressures for the conversion of farmland to other uses, bringing agricultural land into the decision-making processes of the planning system. Social change in the countryside has also brought its own pressures as developers seek to meet the demands of the new rural residents and as more people seek a place in the country. Furthermore, countryside recreation is increasing in importance. The other main source of pressure on rural land comes from industry. The urban-rural shift of manufacturing and services is now long established. It too is constrained by the operation of the planning system, and it takes particular forms in particular places. The uncertainty surrounding agricultural land and buildings has increased pressure on many rural localities to broaden their economic structures.

Through the examination of a variety of development processes we can begin to build up a picture of how particular places come to take on their social and material shapes. These processes may not be discrete either spatially or sectorally, i.e. they may extend across different spaces and may cross sectors, and it will be necessary to follow them in and out of the places that interest us. However, we can seek to illustrate how they give rise to distinctive and (possibly) divergent forms of rurality. Thus, the mode of analysis we propose can account for either increasing homogenization of rural space or its differentiation. We want to use this approach in order to understand the social and material shape of one locality, a locality that appears to be dominated by the middle class. We want to examine how this domination has been achieved. But, more specifically, we ask how do the various processes which underlie the composition of this place give rise to an outcome that can be generalized in class terms?

One way of approaching this question is by considering development as a dynamic interaction between "internal" struggles to promote or resist development (involving a particular constellation of actors) and the wider context of which the development processes are part. Following Mormont (1990), we consider that rural space has a wide variety of uses or values imposed upon it and these arise from the negotiations and struggles between a range of social actors. Where the uses are subject to economic criteria, then market values will (usually) predominate. But where other criteria are imposed, such as those associated with the environment, then alternative values and uses will shape rural space. There is in Mormont's view "no longer one single space, but a multiplicity of social spaces for one and the same geographical area, each of them having its own logic, its own institutions, as well as its own networks of actors" (1990: 34). For Mormont, this leads to a call for a reconstituted rural sociology whose subject "may be defined as the set of processes through which agents construct a vision of the rural suited to their circumstances, define themselves in relation to prevailing social cleavages, and thereby find identity, and through identity make common cause" (41).

We have taken this perspective as our starting point in proposing how studies of rural change might best be undertaken in particular places (Marsden et al. 1993: Ch.6). Briefly, we have characterized rural localities as "meeting places" where particular sets of social relations intersect. We have called these social relations or associations between actors (the collective actors) "networks" and argued that the rôle of analysis is to show how these networks are woven together and how they operate over particular temporal and spatial scales. Within these networks, actors will be bound together in ways which allow them to formulate their interests and act upon them (using the resources derived from network associations). This approach allows us to analyze interest formulation, social action and association (network construction) concurrently. We have suggested that the land-development process can be studied in this way and thus the rural locality can be considered as a series of social and physical outcomes as actors pursue their perceived interests within these networks. Place is thus a meeting point where networks intersect and where some actors may impose their interests over others. Over time these outcomes build upon one another to yield socio-spatial formations. And these formations provide the context for future actions. The simple methodological principles that thus fall out of this approach are, first, as far as possible to "follow the actors" as they formulate interests, make associations and attempt to enrol others to their cause and secondly, to identify how the context constrains and facilitates such action. Furthermore, we might ask how actors, in association with one another, effect changes in the overall context and how these changes provide the preconditions for future action.

While acknowledging that these processes fuel patterns of "uneven development" we also portray rural areas as demonstrating some amount of coherence. We have presented this in the form of four "ideal types" which characterize the range of outcomes we might expect to find to the key economic, social and political processes shaping the English countryside. These are as follows. First, the *preserved countryside*: this is perhaps evident throughout the English lowlands, as well as in attractive and accessible upland areas, and is characterized by anti-development and preservationist attitudes and decision-making. Such concerns are expressed mainly by new social groups in the countryside, such as middle-class fractions (Cloke & Thrift 1990) who may impose their views through the planning system on would-be developers. In addition, demand from these fractions provides the basis for new development activities associated with leisure, industry and residential property. The reconstitution of rurality is often highly contested by articulate consumption interests who use the local political system to protect their positional goods (Hirsch 1978). Secondly, the *contested countryside*: this refers to types of countryside which lie outside the main commuter catchments and may be of no special environmental quality. Here farmers (as landowners) and development interests may be politically dominant and thus are able to push through development proposals. These are increasingly opposed by incomers who adopt the positions that are so effective in the *preserved countryside*. Thus, the development process is marked by increasing conflict between old and new groups. Thirdly, the *paternalistic countryside* refers to areas where large private estates and large farms still dominate and the development process is decisively shaped by established landowners. Many of the large estates and farms may be faced with falling incomes and are thus looking for new sources of income. They will seek out diversification opportunities and are likely to be able to implement these with relatively few problems. They are still likely to take a long-term management view of their property and adopt a traditional paternalistic rôle. These areas are likely to be less subject to external development pressure than either of the above two types. Fourthly, there is the *clientelist countryside*: this is likely to be found in remote rural areas where agriculture and its associated political institutions still hold sway but where farming can be sustained only by state subsidy (such as "less favoured area per capita" payments and welfare transfers). Processes of rural development are dominated by farming, landowning, local capital and state agencies, usually working in close (corporatist) relationships. Farmers will be dependent on systems of direct agricultural support, and any external investment is likely to be dependent upon state aid. Local politics will be dominated by employment concerns and the welfare of the "community".

In this book we examine in detail the first of these types, the *preserved countryside*. Retaining the land-development process as our vantage point, we seek to show how one locality is constituted and reconstituted by struggles around development issues. We analyze in some detail the means by which rural space is moulded in the course of the development process and we examine the effectiveness of the middle class in shaping the new "rurality". In Chapter 1 we sketch in the sociological background to our study, showing how the rôle of the middle class in countryside change has come to be recognized by rural sociology in recent years. Studies have begun to show how middle-class associations or groups are becoming increasingly effective at representing their interests in the rural domain. A key arena in which these representations are made is planning. We therefore examine the ways in which the planning system is used either to maintain or change rural space and show how middle-class groups are seeking to mould the social and physical shape of rural areas into forms which accord with their vision of such places. We wish in this book to examine the extent to which such practices are taking place in one locality, Aylesbury Vale in Buckinghamshire. In Chapter 2 this locality is placed in its regional context, illustrating why the South East of England is a particularly good place to undertake this study. Throughout the 1970s and 1980s the region has become an "engine" of growth for the British economy as a whole. The South East has become dominated by service industries, particularly in the so-called "western crescent" (the counties to the north and west of London) where economic growth has been concentrated in recent years. Economic buoyancy has been accompanied by an enhanced middle-class presence, increasing the dominance of owner-occupied housing and increasingly affecting the rural villages which have become sought-after residential locations. Despite the existence of a preservationist planning system, the patterns of change have been too strong to be contained in growth centres and "boom" towns. But having achieved a place in the country, we argue, many new rural residents are now seeking to use the planning system to "hold the line" between the urban and the rural. Aylesbury Vale is near the centre of these pressures.

People can move to rural areas only if housing becomes available. This occurs if either new homes are built or existing property comes onto the market. In Chapter 3 we examine the housing land-development process and show that a policy framework has evolved to ensure that new housing is built, particularly in areas of the South East, such as Buckinghamshire, where demand for housing has accompanied economic growth. However, this framework must be placed in the context of a generally restrictive local planning policy which seeks to ensure that development is concentrated in towns and growth centres. Thus, rural housing becomes a much sought-after "positional good". One response has been to build housing that embodies rural characteristics, such as green space and a "village atmosphere", on the periphery of towns in order to try and steer the demand for rural housing into "urban" locations. We illustrate this with a case study of just such a development. However, some rural housing has become available and in recent years this has often been as a result of farm building conversions. Barn conversions are the most likely form of housing to become available in up-market, exclusive, truly "rural" villages, where they command a high premium. We therefore examine a case study of converted housing. Lastly, we look at an attempt to provide a new housing estate, including a social housing component, in a highly developed village and at the reaction of local residents to the proposal.

The provision of housing is crucial in determining the development of particular villages. In Chapter 4 we turn to look at three contrasting rural settlements in our study area. One is an estate village where little development has taken place and which has retained both traditional rural features and a traditional (although ageing) population. The second is a "suburban village", close to the county town of Aylesbury, which has all but lost its village identity, as it has been subject to high levels of housebuilding during the 1970s and 1980s. The third is a medium-size village with a mixed population which has in recent years seen some up-market barn conversions. One part of the village (the "heart") has become dominated by middle-class incomers who now seek to define community life. We compare these places to show how development processes lead to different outcomes at different times. These outcomes provide the "preconditions" for new rounds of development (or perhaps resistance to development).

The village studies show how, in one locality, a range of different settlement types and social formations may be present. We must therefore exercise caution before drawing broad generalizations about the social character of localities. However, we still retain the view that there is an increasing middle-class presence in Aylesbury Vale, even though this is spatially variable. In the following chapters we examine whether this class is increasing its influence over land uses in the open countryside. In Chapter 5 we turn to examine the recent crisis of agriculture and the way this has manifested itself locally. With problems of surplus food, budget-

ary pressures, and public awareness of increasing environmental damage, farmers have seen their industry beset by great uncertainty. This has
led some to exploit their assets as fully as possible *within* agriculture – we
present some case studies of this type in Chapter 5 – while others have
sought to exploit them *outside* agriculture. In Chapters 6, 7 and 8 we
present studies of land development in this latter category. In Chapter 6
we see farmers putting land over to golf courses. We argue that this new
use for the countryside is acceptable in areas such as Buckinghamshire
because it largely accords with middle-class conceptions of "green
space". However, in Chapters 7 and 8 we look at proposed new uses for
agricultural land which are vehemently opposed by rural residents: quarrying and landfill. We show in some detail how local residents come
together to form action groups to oppose these developments.

The crisis in agriculture has resulted not just in surplus land but also
surplus buildings. In Chapter 9 we examine how farmers in Aylesbury
Vale have sought novel uses for their buildings, in this case showing how
they have gone from using buildings for agriculture to establishing small
industrial estates. These developments were not subject to the kinds of
objections faced by the minerals and landfill companies, but eventually
fell foul of the planning system as planners came to be concerned about
industry moving into the countryside. The tone of the planning authority's attitude to such developments on farms shifted from encouragement
(i.e. the perceived need to diversify the rural economy) to hostility (i.e. the
belief that these developments are not in keeping with the aesthetic character of the countryside). Here, middle-class values were not asserted by
action groups but came to be embodied in the plans that guide local-
authority decision-making. This was perhaps a more disguised but effective route to the maintenance of an "exclusive" countryside.

The action-in-context approach we adopt in this study allows us to
examine a multitude of processes and ways in which they shape the social
and material character of a particular place. The "rural" is "made". It is
the outcome of a variety of processes which cross-cut the locale. In the
first volume of this series (Marsden et al. 1993) we outlined in detail how
our approach focuses upon the study of actions within the processes of
land development. We argued that a particularly appropriate method is
the case study. In the present book we make extensive use of this device.
The studies presented here were undertaken from 1989 to 1992 and con-
sist of in-depth interviews with the key actors in specific development
projects. These interviews were supplemented by the collection of documentary evidence. Although the case study approach is useful because it
allows analysis of social *processes*, it suffers from problem of representativeness: that is, how can one case be taken as indicative of general
trends? We have tried to overcome this weakness in several ways. First,
in presenting the case studies we make continual reference to the context

of the cases and the ways in which this general context is both illuminated and (occasionally) modified by the outcomes of the studies. Secondly, we supplement our case studies by other sources of data, notably a series of in-depth interviews in three villages (see Ch. 4), in an attempt to identify some of the social impacts of development, and two social surveys. The latter were, first, a stratified random sample of householders across the county of Buckinghamshire (2,000 sample – including respondents from seven villages) interviewed as part of Buckinghamshire County Council's Structure Plan Review process in 1991. They were questioned on their attitudes to planning and development. Secondly, we undertook a longitudinal study of farmers in Aylesbury Vale (300 sample) involving questionnaire surveys over a three-year period (1989–91) and in-depth panel interviews (70 sample) over the same period. This survey was sponsored by the EC and was concerned with collecting evidence on (a) land-development strategies (b) financial and credit links and (c) policy uptake and assessment. In the following chapters we use these sources interchangeably. In Chapter 2 the Buckinghamshire County Council Survey is used to exemplify social change in the locality; in Chapter 5 the farm survey data is presented to show the nature of agricultural restructuring in Aylesbury Vale. In Chapters 3–9 case studies provide the main source of data (for more detail on the various methodologies, see Appendix 1 and for the main case study sites see Fig. I).

The study of land development across a wide range of sectors, through the analysis of this empirical material, gives us access to both variation and coherence in the key processes currently moulding rural space. We use the locality approach as an empirical, as well as an analytical, tool in order to uncover the social and the material dimensions of rural change. Development takes "concrete" forms, but these forms represent social uses and values. In each type of development we examine how these uses and values differ. However, it is instructive to investigate whether those that predominate can be woven into a coherent account of social and material change, one which allows us to speak of patterns extending beyond our locality. In order to achieve this we attempt to hold the "micro" and the "macro" together through the examination of action-in-context, most notably *class* action and *class* context. To achieve this we seek to provide in the following pages a "balanced" account, one which stresses the contingent nature of situated action but which weaves into this a larger story, one which speaks of much more than land development in rural Buckinghamshire. This issue of "balance" is examined in the concluding chapter where we assess the extent to which restrictions on certain forms of development, and the facilitation of others, are leading to a reconstitution of rurality and whether this can be adequately considered to be part and parcel of the processes of middle-class formation in late twentieth-century England.

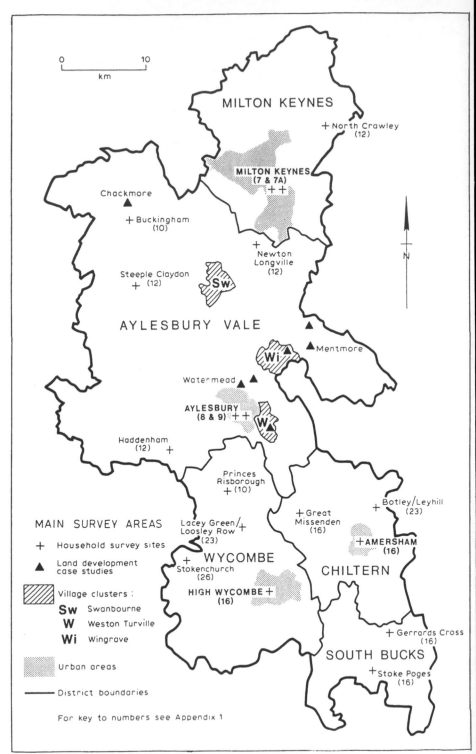

Figure I The study area

CHAPTER 1

Class action and the reconstitution of rural space

Introduction

In selecting Aylesbury Vale as our geographical area of study, we consciously followed much recent research which has demonstrated that the outer South East of England can be conceptualized as a "middle-class space". We are conscious of our debt to this earlier work and we begin this chapter by reviewing some of the recent literature on class analysis, particularly in relation to the middle class. We then go on to show how rural sociology has traditionally focused on the issue of class. We adopt a perspective that focuses on class *formation* and this leads us to consider class *action* and, in particular, how the planning system has become a key arena in which much of this action in the rural domain is played out. The account of class formation we present here provides the starting point for our analysis of the land-development process in later chapters. We wish to demonstrate that classes are not fully formed entities determining everything that comes within their ambit; rather they are continually "made" and reproduced in a variety of different arenas through processes of struggle, conflict, persuasion and enrolment. Furthermore, such processes are place-specific; that is, class formation is, at one and the same time, *place* formation. In the following chapters we provide evidence from a variety of sources and perspectives to show just how complex the processes of formation and reproduction can be.

However, we begin by considering why class analysis has seemingly fallen out of favour in rural studies in the UK. Ostensibly, the aversion to notions of class in rural studies may reflect rural ideology which Cloke & Thrift believe "traditionally presents the countryside as an essentially classless society even if an unequal and hierarchical one" (1990: 165). This ideology has in turn been bolstered by the "invisibility" of rural issues in mainstream sociological debates about class (Hamilton 1990: 229). Despite the influence of these views, there is a recognition, amongst at least some practitioners of rural studies, that class has been unduly neglected: Paul Cloke, in a review of rural geography in Britain, has argued that "changing class structures, particularly the infiltration of different fractions of the middle class ... [are one] of the major issues requiring serious attention" (Cloke 1989: 191; see also Cloke & Moseley 1990). As yet class analysis still seems to be of secondary concern to many sociologists and geographers working in rural studies (although see Cloke & Thrift 1987, 1990). In one sense this is surprising given the interest in such topics as restructuring, gentrification, deprivation, regulation, etc., which depend, at least in part, on conceptions of class for their explanatory value. Yet, for some reason, class analysis *per se* is absent. In part, the neglect of class in rural studies can be traced to problems associated with the concept in mainstream sociology where, at the very least, considerable confusion surrounds the term. In the first section, a short review of contemporary debates within sociology shows how class analysis might usefully be reconstituted in the rural domain. The bulk of this discussion is taken up with a summary of a recent book by Savage et al. (1992) which provides an account of middle-class formation in contemporary Britain. In the second section we illustrate how class analysis can be used to reinterpret many of the most influential trends in rural research, using the middle class as a focal point. We then turn to consider what we believe to be one of the key arenas in which processes of class formation are played out in the rural domain: planning.

Conceptualizing class and class formation

While the notion of class is used by many different people both to explain and describe social phenomena (despite recent attempts to herald the arrival of a "classless society"), for sociologists there seems any amount of difficulty associated with the use of the term. These difficulties include problems of definition, such as where does one class begin and another end (the boundary problem), who should be included and who should not, and, even more vexingly perhaps, whether classes can in any sense be seen as coherent and conscious social actors. Through a brief survey of some contemporary debates surrounding the concept of class, we can begin to understand how these difficulties have arisen and how they might be overcome.

In a useful summary of class analysis, Crompton (1993) identifies two main approaches which have tended to diverge in recent years. First, class analysis can be used to *describe* levels of social and material inequality. Here classes "summarize the outcome, in material terms, of the competition for resources in capitalist market economies" (10). A long tradition of empirical investigation into class *structure* has thus developed, using large data sets derived from social survey work or using aggregates of employment and occupations (most notable here would be the work of Goldthorpe and colleagues, e.g. Goldthorpe et al. 1980, Goldthorpe & Payne 1986). On the other hand, there is a tradition that sees classes as corresponding to the structures and power relations which have *produced* these inequalities. Here the emphasis is placed on the processes of occupational and class structuring which cannot be adequately analyzed through aggregates of occupations. Classes must be treated as social collectivities investigated in particular social contexts; they must "be studied in relation to the institutions and organizations which articulate their claims – trade unions, political parties, and so on" (Crompton 1993: 113). The preferred mode of analysis is some form of case study. The most notable example here is E. P. Thompson's *Making of the English working class*, an historical study of class *formation*.

Resting upon these differences, Crompton discerns a further set of distinctions, the most important of which is that between structure and action. Those undertaking work on class structure (such as Goldthorpe and Wright) tend towards an analytical separation between structure and action, while those incorporating action treat structure and action as some kind of unity. This distinction has resulted in heated debate over the status of the two approaches, casting a shadow over the conduct of class analysis itself. As Savage et al. (1992: 220) note, the quantitative tradition of mapping class structure through survey data has become dominant in the past ten years, yet a growing band of critics argue that simply identifying the existence of classes goes no way to explaining how they came into existence or how class categories help in analyzing any form of social action (Hindess 1987, Lockwood 1981).

Pahl (1989) crystallizes the problem of examining class action, which he claims can be summed up by the "mantra" of "structure–consciousness–action" (SCA). The SCA chain refers to the relationship between the social and economic circumstances in which actors are embedded (i.e. structure), consciousness of those circumstances, and the resulting action, which (may) seek to transform those circumstances. Pahl argues that the "links in the chain are rarely seen as problematic" (711) but they cannot be considered by simply allocating individuals to class categories via survey data. This type of work remains transfixed by structure and provides little understanding of how consciousness and action come into being. He says "scholars have behaved *as if* the links in the SCA chain were self-evident. Once it is understood that such links have yet to be discovered an important programme of research designed to advance theoretical understanding can be opened up" (719). The task of class analysis is to discern the links in the chain. In response, Goldthorpe & Marshall (1992: 385) have argued that there is no reason why individuals sharing similar positions in the class structure will "automatically" develop a shared consciousness and act in concert. Pahl (1993) believes this demonstrates that class analysis has conceded too much theoretical ground and simply cannot explain why consciousness and action *do* come into being.

Another participant in this debate believes that empirical class analysis using longitudinal survey data is not well suited to understanding the processes of economic and social transformation. Mullins (1991) argues that sociologists have consistently neglected *development* and thus have little understanding of the dynamics of social structures or the impact of social forces. In his view, this neglect is tied to the discipline's static methodology which only allows "snapshots" of the social structure. However:

> An understanding of development . . . is predicated on an understanding of the social forces involved in this development; forces like class relations and consumption relations. This is because development results from the combined impact these forces have on social structures and for this reason, class analysis and consumption analysis are of value only if – in the end – they show whether class and consumption contribute to these transformations. (Mullins 1991: 119)

Leaving the question of consumption to one side, the form of class analysis being proposed here is dynamic, and it seeks to emphasize the way in which consciousness and action come to be implicated in processes of transformation. As we move away from some overdetermining class structure, issues of collective action and class formation come to the fore.

Commentators such as Barnes (1990) believe sociological theory is increasingly coming to recognize that not only the transformation but the maintenance of any given social order results from collective action, yet the study of such action has been relatively neglected. While sociological analysis is using "class structure" as an explanation, little attention tends to be paid to how such a structure is "made"; the task is simply that of mapping the empirical data in a form that accords with the structural parameters. Once we shift our focus to see how this structure arises from social action, then we need to consider the variety of forms in which actors come together and act collectively. In Pahl's view one consequence of studying collective action in all its forms might be a more sophisticated understanding of the relationship between class action and class structure:

> From informal social networks through families, kinship links and the whole range of formal and informal associations of civil society people are engaging in voluntary solidaristic and collective activity for a variety of goals . . . . Overwhelmingly these social groups are issue-oriented and territorially based . . . . whether these informal associations are the stuff out of which the links in the chain [SCA] can be forged is hard to say. At present work on such associations,

locality-based groups and networks is not directly focused to the theoretical lacunae in the SCA model. However, there is no reason why this should not be done. (Pahl 1989: 719)

Likewise, Savage et al. (1992: Appendix 1) examine the contemporary debates surrounding class and conclude that the most useful way forward is to adopt an approach which sees social classes as collectivities and which seeks to demonstrate whether they have an impact on processes of historical change. The authors note that it is "difficult to integrate a theory of class based on a synchronic examination of class positions into an account of diachronic historical change" (p. 227), but the "key point" is that the separation of synchrony and diachrony may actually prevent any adequate analysis of the historical dynamics of class formation being developed. As soon as attention is focused on historical processes, it becomes impossible to specify distinct class positions. As society changes, absolute measures of class and class position also change; thus, "attention is necessarily refocused away from the positions themselves to the processes that structure these class positions" (p. 228). Savage et al. prefer an approach (derived from Przeworski 1977) that sees the process of class formation as the continual (re)organization of class positions. So the latter (those elements which are captured through quantitative analysis) can be conceptualized as the outcome of the processes of formation.

Social classes are collectivities, that is groups with shared levels of income, life-styles, cultures, and political orientations. These collectivities, in order to qualify as classes, must have their "roots" in processes of exploitation. The concept of exploitation specifies a relationship between classes, i.e. how one social class distinguishes itself from another and how it maintains that distinction. If we seek to understand the maintenance of such difference, our attention shifts to the ways that classes and other collectivities "store" their advantages. Any gains made by a collectivity must be stored if it is to remain stable over time. Furthermore, these gains must be transmitted to other members (e.g. offspring) to perpetuate advantages. Such gains and advantages are considered as forms of capital, and two such forms are usually identified: economic and cultural. The first has traditionally been regarded as the main axis of class formation and exploitation (in both Weberian and Marxist accounts), and it derives from relationships in the productive or market spheres. The second is theorized by Bourdieu (1984), and utilized by Savage et al., as a crucial component of class formation. Cultural capital has to be created and legitimated and is used to distinguish groups or classes from one another. It takes the form of "intellectual capital", which is transmitted via educational standards, and "embodied capital", which is transmitted to offspring through standards of health care, patterns of adornment, fashion, bearing, etc. These components of cultural capital can also be used to acquire professional credentials and skills that may lead to the acquisition of capital in the economic sphere. The latter entails the deployment of cultural capital in particular organizational contexts in order to allow material rewards – high income, job security, etc. – to be gained.

Savage et al. argue that capital is rendered into assets and that these can be utilized to aid class formation in certain contexts. They identify three types: cultural assets, property assets and organizational assets. These assets relate to one another in complex ways and should not be regarded as symmetrical; they vary in part according to the social circumstances in which they are employed. In actual fact, the authors argue, cultural assets and property assets are becoming increasingly important to patterns of middle-class formation while organizational assets, which have always been problematic for storing advantage (they tend to be tied to specific organizational contexts), are becoming less useful. A shift in the dominant organizational form of the advanced capitalist economies from large bureaucratic ("Fordist") hierarchies to decentralized ("post-Fordist") networks is reducing the provenance of organizational careers via internal labour markets and increasing the rôle of self-employed or professional specialists. As a result, "the pyramidical organizational hierarchy is disrupted, and the power of organization assets alone to convey reward is severely questioned. Firms increasingly look to those with specific skills to perform particular jobs, rather than relying automatically on bureaucratic procedures" (p. 66). The authors argue that cultural assets become encoded in expert knowledge; this knowledge is internalized in its individual practitioners while organizational knowledge becomes tied to particular organizational forms and contexts. The former is therefore mobile; the latter is relatively fixed.

Savage et al. go on to consider a particularly distinctive feature of social change in post-war Britain – the rise in owner-occupied housing. They point to an increase in the proportion of the population owning their homes, from 10% in 1914 to 70% in 1990 (pp. 80–82), so that by the end of this period virtually the whole of the middle class had purchased their dwellings. Moreover, the significance of owner-occupation and its rôle in middle-class formation is heightened by associated factors, such as house price inflation which has been endemic: since 1943 house prices have risen five-fold (p. 85), with middle-class occupiers achieving amongst the highest capital gains (see Forrest & Murie 1990). Property assets, which have hitherto been of relatively marginal importance outside the ranks of the petty bourgeoisie, are becoming more integrally tied to processes of middle-class formation. Savage et al. suggest that "increasing numbers of the middle class can draw upon both property and cultural assets" (p. 59).

Class formation, they point out, "does not take place on the head of a pin" (p. 33); it takes place in specific spatial contexts. These contexts are partly shaped by changes in the employment structure which are both caused by, and are a response to, middle-class spatial formation. Spatial mobility is a key feature of the middle-class lifestyle and this has the consequence of allowing middle-class households gradually to acquire more prestigious properties in the course of housing *careers*. Thus, "those areas with the highest proportion of professionals and managers tended to have the highest house prices in 1985" (p. 88). Property assets, in the form of owner-occupied housing, become central to class formation. But more than this there is a "trend towards the investment of cultural assets in housing . . . so that the aesthetics of the middle-class residence plays a major part in the exhibition of specific cultural tastes and values" (p. 94). Developers, moreover, have responded to these new demands by catering for specific middle-class housing tastes. Cultural assets are not simply enshrined in the house but also in the surrounding area: thus "specific housing areas are taking on particular rôles, geared to differing household types" (97), often described as "gentrification" in reference to the way in which middle-class "colonizers" or "incomers" convert residential properties to their own tastes. At the neighbourhood level this process changes both the social and material complexion of an area, marking it off from working-class areas. The South East of England is identified by Savage et al. as the region in which the economic and cultural formation of the middle class is most advanced. It is expressed culturally in patterns of consumption and types of neighbourhoods.

Finally, Savage et al. examine middle-class politics. The main focus here is voting patterns, but they do make some comments about other forms of middle-class political action most notably associated with "conservation groups". They draw upon work by Short et al. (1986) in Berkshire on the rôle of middle-class owner-occupiers in preventing unwanted developments in the county, which they characterize as a "defensive politics designed to preserve their privileged position" (Savage et al. 1992: 209; this is discussed in more detail in the next section).

Given the sophistication adopted by the Savage et al. theoretical approach, it is a pity that their analysis of middle-class formation is not carried more fully into the portrayal of culture and politics. This may be because they lack the empirical work necessary for an examination of the "micro-processes" of class formation in particular places at particular times; their empirical evidence is largely drawn from survey-based material. However, the account presented in *Property, bureaucracy and culture* demonstrates the multifaceted nature of class formation. Once we begin to move away from a conception of class which sees it as "grounded" in the sphere of production then we can begin to appreciate that classes become defined in *multiple* arenas simultaneously. One of the problems with *Property, bureaucracy and culture* is that little evidence is presented of middle-class formation in the workplace apart from some general comments on industrial and organizational restructuring. However, the value of the work lies in the way it illustrates how class formation takes place simultaneously in the economic, cultural and political spheres. Moreover, the *spatial* implications of middle-class dominance are addressed and it is here that the implications for the present book become clear, for we will argue that rural areas come to play a distinctive rôle in the process of class formation and are in turn rendered into a particular class shape.

In concluding this section we should be clear about the form of class analysis that is being proposed here. The problems associated with structuralist conceptions of class – such as the sheer difficulty of moving from structure to consciousness to action, and the static nature of the prevailing methodologies – have led some researchers to seek out more dynamic forms of analysis. Savage et al. focus on class formation and usefully consider a range of different domains in which this occurs. Classes are considered to be collectivities with shared resources and similar cultural and political orientations. If we are concerned with class formation then we must understand how these collectivities come into being and distinguish themselves from others. In this regard it is worth referring to recent work by Eder (1993) who argues that we should begin with collective action and then ask how the (re)production of class difference is connected to such action. Thus, "class comes last: it is the most restrictive variable; it reduces the variability of events that explain what is going on in reality" (Eder 1993: 10). Class "structures" the processes of collective action; that is, class is used as a means of identifying how resources, attitudes and attributes are distributed and how these come to be used in the action process. But the action process will in turn lead to a stabilization, consolidation, redistribution or redefinition of these resources, attitudes and attributes. Thus, it is perhaps most appropriate to put class at the end of the analysis; it becomes an *outcome* of collective action and these outcomes provide a *structure* of enablement and constraint for future action. In studying how rural Britain is both reproduced and transformed we can therefore focus on how and why individuals are engaging in collective action, the types and levels of resources, assets and attributes available to them, the different issues and territories around which they mobilize, and the outcomes, which can be characterized in terms of class formation and class space.

Class action and the reconstitution of rural space

Despite our earlier criticism that class analysis had become relatively marginal to rural sociology and geography, this does not mean that it has been altogether absent from the rural studies agenda. In fact it is now becoming commonplace to assert that the rural is "middle-class territory"

(Buller & Lowe 1990: 27), a view which Phillips believes is becoming an "emergent orthodoxy" (1993: 131). However, little attention has been paid to middle-class formation in rural areas or to the rôle of rural space in the "making" of the middle class. Nevertheless some of the most influential post-war analyses of rural Britain have used conceptions of class as part of their explanatory frameworks. In this section we shall briefly review certain key texts.

First, however, consider this quote: "The city, our great modern form, is soft, amenable to a dazzling and libidinous variety of lives, dreams and interpretations. But the very plastic qualities which make the city the great liberator of human identity also cause it to be especially vulnerable to psychosis and totalitarian nightmare" (Raban 1974: 8). This quote from Raban's book *Soft city* allows us to make two observations. First, space is malleable, plastic, "soft". While Raban seems to believe that this is the exclusive provenance of urban space (his comments on small communities include words such "upright" and "censorious" – 223) we will focus here upon the reconstitution of rural space. This may not be *as* soft and plastic as the urban but this does not mean it is static and unbending. As we shall see, rural space has proved itself malleable to particular dominant social groups.

The other observation we can draw from Raban is the reference to the "dark side of the city" as he also says "to live in a city is to live in a community of people who are strangers to each other . . ." (7). The city is a place of great uncertainty and instability where individuals may find extreme difficulty in achieving any real sense of security. Raban paints a bleak picture of "psychosis" and "nightmare". The provenance of such a description of urban life might indicate why certain social groups have come to seek out other spaces. The first of the twentieth century middle-class habitats was suburbia, which Fishman (1987) believes expresses a middle-class vision which frees them from the corruption of the city and sees them "restored to harmony with nature, endowed with wealth and independence yet protected by a close-knit stable community". Suburbia is "the collective creation of the Anglo-American middle class: the bourgeois utopia" (x). However, as more people have moved from the metropolitan areas into the countryside so the values which were asserted in suburbia came to be found in new (primarily village) contexts. Val Williams neatly summarizes the virtues represented by the English village in the following way:

> Being part of a village is like being a member of an extended family and city dwellers, often far away from their roots, long for this sense of belonging. Viewed from the city, village life seems so secure, with its bastions of the vicar, the schoolteacher, and the local squire. The school is a cosy, well disciplined place where the

9

old values persist, a log fire burns in the pub and the troubles of the workplace and the home can be forgotten in its flickering ambience. (1993: 34)

The study of social actors in pursuit of such aspirations has been a recurring theme in rural sociology over the past thirty years.

Ironically perhaps, given his central rôle in disputing the utility of class analysis, the most appropriate starting point is the work of Pahl who conducted a series of village studies in Hertfordshire approximately 30 miles from London. Most post-war rural community studies (as Pahl notes, 1965: 5) had been undertaken in traditional settlements in peripheral areas relatively untouched by the "pernicious" effects of urbanism. Pahl, on the other hand, wished to examine the urban nature of rural areas, particularly in what he calls the "metropolitan" villages of South East England. In this context, he asks "where *is* the city and what is its effect?" (1970: 270, emphasis in the original).

For Pahl, a village is best understood as a "state of mind". In saying this he seeks to focus attention on how villages are created as communities. He notes that rural people, i.e. those who are bound to particular areas by their spatial immobility, acquire something of an urban outlook (through mass media and improved transport links) while urban people, i.e. those previously resident in urban areas, bring "arcadian" visions of country life into the village (1965: 6). The influence of this incoming middle-class population is, Pahl argues, a post-war phenomenon; in his study 81% of incomers had arrived in the period between 1945 and 1961 (1965: 9). The middle class has a high degree of mobility which allows it to choose places in which to live. Furthermore, this mobility also allows the middle-class incomers to live within the village but conduct much of their lives outside it: work, friendship networks, leisure activities and shopping may all take place elsewhere. The attraction of the village lies in the physical surroundings and the pattern of social relationships which are likely to be found there: "Unlike, say, the suburbs, the village situation involves interaction with other status groups" (1970: 274). This interaction is likely to take place within what is imagined to be a "real community". However, Pahl argues,

The middle-class people come into rural areas in search of a meaningful community and by their presence help to destroy whatever community was there. That is not to say that the middle-class people change or influence the working class. *They simply make them aware of national class divisions thus polarising the local society.* (1965: 18, emphasis in the original).

This conclusion is drawn by recourse to the idea that there is an alternative village "in the mind". Pahl argues that "part of the basis of the

10

village was the sharing of deprivations due to the isolation of country life". He believes that "middle-class people try to get the 'cosiness' of village life, without suffering any of the deprivations" (ibid). The middle and working classes move "in separate worlds" and view each other "with only partial understanding" (1965: 11&18).

Pahl sees the middle class as "cosmopolitans" and the working class as "locals". Having entered the village the cosmopolitans attempt to put down localistic roots and enter the "spirit" of the community which "almost without exception" they describe as "friendly". They are also opposed to new development "which is inevitably out of 'character'" and they thus become the "self-appointed guardians of tradition and rusticity" viewing the (local) manual workers as "props on the rustic stage" (1970: 274). This is what defines the character of the village and seems to mark off rural from urban communities, for only in such villages are "groups which, in the 'normal' urban situation, would be socially distant . . . forced into an unusual consciousness of each other" (1970: 275). The influx of new social groups into the village thus helps to "crystallize a class situation" (1970: 277 fn) and what is now sociologically interesting about such communities becomes the extent to which "national" influences (such as class) come to shape the character of the "local". Thus, Pahl concludes "the rural sociologists main concern, as I see it, is to explore the impact of the national on the local" (1970: 294).

While this work can be read as one of the earliest and most influential post-war investigations of middle-class formation in a particular place, demonstrating how rural space becomes remoulded in the process of social change, it also illustrates the way in which a conception of the *national class structure* is used to "explain" social outcomes in a particular place through the assumption that this structure could be unproblematically linked to consciousness and action. The national defines the local.

Some of the themes identified by Pahl as worthy of further investigation were taken up by Newby and his colleagues in their studies of the social structure of rural East Anglia (Newby et al. 1978, Saunders et al. 1978, Newby 1980a, 1980b). This work covered many aspects of rural society and politics and not all of these need to be rehearsed here. However, the influence of the "national" on the "local" and the relationship between "locals" and "incomers" were recurring themes and it is worth considering their handling of these.

Newby et al. (1978) focus mainly on capitalist farmers, their place in the national class structure and how they attempt to maintain their dominance in the rural locality of East Anglia. However, in the course of their study the authors discover a "tremendous feeling within East Anglian villages of having been overrun by an invasion of outsiders" (194). The reaction of agricultural workers, farmers and landowners to the incursion of outsiders is to retreat into two tightly knit communities: one is the "encapsulated community", consisting of "true country people who really understand agriculture" and "respect local farmers and landowners"; the second is the "farm-centred community" which "revolves around the farm itself where farmers and farm workers both live 'in' . . ." (194). The creation of these communities within particular villages has two main effects: first, farmers and farm workers are brought into very close proximity to one another, increasing, the authors believe, the control farmers have over their workers (often bolstered, it is argued, by notions such as "localism" and "community" which "mystify" the essential class division between employers and employees – see Saunders et al. 1978); secondly, divisions between "locals" and "newcomers" are (re)produced and clearly marked, often resulting in some amount of conflict between the two groups.

Newby et al. portray a farming community closing in on itself using, for instance, survey evidence which purports to show that farmers' social networks are extremely confined. In general, they conclude, farmers "simply do not have non-rural, non-local friends . . . they rarely meet socially people from outside the countryside or even from outside agriculture and agriculturally related industries" (210). Newcomers, however, bring to the village "urban" styles of living and frequently clash with the farming community over issues connected with access to land and protection of the environment. However, the relationship between farmers and middle-class incomers is not simply one of antagonism. In their investigation of local politics Newby et al. establish that farmers and landowners still dominate local political bodies and use their dominance to pursue "conservationist" policies. As a consequence of these policies industry is excluded from the region, for this would compete with agriculture for labour. The result of its exclusion is the perpetuation of a low wage economy and a curb on the in-migration of other members of the working class. Newby et al. argue that the middle class do not challenge the political dominance of the traditional elite because they also benefit from the conservation policies: "both farmers and newcomers desire to preserve the countryside as 'truly rural'. In this sense, all property owners . . . are more or less on the same side of the conservationist fence, with all that that implies for the low wage economy of rural areas" (248–9).

The interpretation that Newby et al. make of these developments derives from their exploration of the "national" class structure in a particular locality. For instance, a particularly vexing question is why farmworkers continually side with farmers whom (objectively) they ought to oppose. The authors believe the "ideology" of community provides an explanation. Thus:

> there is an alliance between the locals of whatever class against the newcomers. Put another way, there is evidence of strong vertical

identification between the traditional lower and upper classes of rural society, this being expressed in such diverse forms as the image of community which they both share, or the gentlemanly ethic to which they both subscribe (313).

Newby et al. consider, therefore, that the movement of middle-class residents into the countryside is consolidating traditional sets of power relations. The outcome is a "conserved" rural space which excludes certain forms of development (industrialization) and certain social groups (the industrial working class). Thus, a rather traditional type of rurality is retained.

This is a different perspective to that presented by Pahl (it is after all a study of a quite different locality at a different time) but it still seeks to show how the social and material shape of the "rural" can be analyzed using a conception of the "national class structure". Again, the concern is not with class formation (the middle class is really just a shadowy presence in this work) but, following Pahl, with how the national class structure can be interpreted in a local area (Newby makes this explicit elsewhere –1980b: 240).

The local studies conducted by Pahl and Newby et al. tended to play down the specificity of local social processes and the social distinctiveness of the respective localities (Bradley & Lowe 1984: 5). They were concerned to show how the *national* class structure played itself out in a rural context. For our purposes what is interesting about these studies, however, is the extent to which they show *variation* in class formation between particular places at particular times.

The uneven formation of class, and the rôle of the middle class in shaping rural space, is further highlighted by Barlow & Savage (1986) in their analysis of development in Berkshire during the early 1980s. Unlike Newby et al. they find that the middle class and farmers are in conflict. They note that the conservationist lobby has emerged as a powerful force during a phase of sustained economic growth in the county. Many of the most active participants in this lobby, they argue, "are people who moved to a 'rural' area under the impression that their development was the last which would be permitted, only to find house-builders . . . pressing for more land release" (170); they thus wish to prevent any further development in their immediate locality. Farmers and landowners, on the other hand, are seeking to take advantage of the buoyant land market to cash in on the potential development gains. Barlow and Savage thus note a "growing disharmony" (177) between the conservation lobby and its middle-class supporters, and farmers and landowners.

The activities of a politically active middle class in this buoyant economic locality are also referred to by Short et al. (1986) in their study of planning and development in Berkshire during roughly the same time period. They note an increase in the number of middle-class conservation and residential groups since the 1960s, mainly concerned with preventing further development in (certain areas of) the county. These groups, particularly those associated with the residents, tend to be parochial, being concerned with the physical and social aspects of their immediate environment. Short et al. record this phenomenon as being particularly strong in villages where "there is a powerful ideology which sees in a village location the hope of reasserting a moral arcadia away from the anonymity of mass urban society" (209). Those most concerned to protect the villages from development are, again, the most recent residents and not all villagers are involved – "it is mainly the middle-class and upper-income residents" (209) – and they conclude that these "no-growth" groups have been influential in creating an articulate and powerful lobby "which has sensitized local members [of planning committees] to the issues of growth containment and growth deflection" (238).

What these studies begin to illustrate are the general principles of class formation laid out by Savage et al. (1992). As certain local economies become increasingly characterized by middle-class occupations so there is an increased middle-class presence in proximate rural localities. Once class members begin to establish themselves in these areas they begin to actively mould (politically and culturally) the social and material shape of the locale, often attempting to reproduce particular conceptions of the rural (Thrift 1989). This type of activity is likely to make these areas even more attractive to those who wish to fully embrace a middle-class lifestyle, but is also likely to increase tensions between such incomers and more longstanding rural residents.

Cloke & Thrift (1990), however, believe that a characterization of class conflict which simply pits the middle class (usually described as "incomers") against either the working class or a landowning class neglects the extent to which the middle class itself is "fractured". They suggest that there exists a series of complex intra-class divisions that cut across traditional class divides. Therefore, in different rural localities we should expect to find different "fractions" exhibiting their own characteristic behaviour patterns. In an examination of the middle class (they prefer the term "service-class") they discern five fault lines – public/private sector, gender, life cycle, consumption practices and types of locality – which "provide the basis for many different service-class fractions" (176). Conflict in rural areas, they argue, can erupt around these divisions. They conclude by proposing a series of topics for research into these intra-class conflicts, including: how class fractions relate to broader conceptions of class; how they are spatially dispersed and internally organized; the nature of conflict between the various fractions; how they articulate with other classes; and how far they are able to change the social complexion of various localities (see 177–8).

14

Such a consideration of how the middle class might be "cross-cut" by other divisions alerts us to the complexity of contemporary class analysis, particularly if we wish to avoid sidelining other sources of identity and oppression. However, terming these "intra-class fractions" runs the risk of portraying class as always fully formed and somehow all-encompassing. Class seems to capture these divisions while at the same time being riven by them. If we shift the emphasis to class as both a *medium* and *outcome* of *various* social processes then we can begin to consider how collective action – in many different circumstances and for many different reasons – might contribute to, or indeed undermine, class formation. This is necessarily a dynamic process, and classes are always only stabilized provisionally. It is also spatially uneven. Thus, rural space becomes differentially bound up in an endless process of class (re)composition. Rural gentrification can then be reconsidered as an intrinsic part of class formation. Rural areas, considered as particular ensembles of houses, neighbourhoods and locales, come to play a key role in the spatial configuration of the middle class. Thus, rurality can be also be seen as an outcome of processes of class formation as individuals and collectivities attempt to mould rural space into forms which reflect and perpetuate class identity and difference.

Other facets of the middle-class incursion into rural areas have also been considered in a series of publications concerned with "gentrification" (e.g. Little 1987, Cloke & Little 1990, Cloke et al. 1989). Phillips (1993: 124) notes that the term gentrification is most usually employed to signify middle-class colonization of an area once dominated by the working class. However, in his own study of gentrification in the Gower peninsula he concludes that gentrifiers cannot simply be categorized as middle class. Phillips finds that rural gentrification "may well be, at least in part, the result of the perpetuation of patriarchal gender identities and associations" (138). This suggests that simply portraying rural areas as "middle-class territory" is too simplistic a view of contemporary trends, and indicates that class formation may be a spatially uneven process.

While this type of work yields considerable insight into the ways in which middle-class colonization unfolds, these studies rarely address the issue of class formation head on. There is still a tendency to "read off" the formation of classes from the economic context, as if change in this arena somehow gives other (cultural and political) processes their "underlying logic". In a useful corrective to this view Thrift (1987b) proposes that economic change is itself driven by cultural factors. Thrift argues that members of the middle class have, in general, a predilection for rural environments (again Thrift uses the term "service-class" but for the sake of consistency we will continue with "middle class" – for a consideration of both categories see Abercrombie & Urry 1983). They are, furthermore, able to do something about this predilection: they can either attempt to

15

keep their neighbourhoods as rural as possible or they can colonize rural areas with particular housing and employment opportunities (thus leading to gentrification). As the literature reviewed above indicates, this latter strategy has been pursued in much of rural Britain, particularly in the south of the country. But as Thrift notes, this is not a simple change in the social structure of rural localities, for the process is cumulative: "As more service-class members arrive in an area so work and consumption opportunities, communications, and housing will follow" (79). This is because the middle class does not just follow the location of industry; it also acts strongly to influence location in the first place. There are a variety of reasons for this, including: the rôle of managers and professionals in making location decisions; the need for employers to seek out pools of middle-class labour; and the way middle-class neighbourhoods attract particular services and thus more service sector employees (79).

Although he only provides a thin sketch of this process, Thrift gives an indication of how class structure and class formation can be considered concurrently. The formation of middle-class occupational places in a given locale becomes intimately bound up with other facets of class formation, such as the search for "idyllic" rural living environments. Furthermore, the spatial and social become here intimately bound up in a variety of processes which go into the "making" of a class. Again, class can be considered as an outcome rather than some underlying, determining structure.

The rôle of politics in middle-class formation is indicated in work by Jo Little (1987). Echoing some of the points made by Newby et al. (1978), Little believes that the increased middle-class composition of rural villages is actively encouraged by planning policies which are concerned with environmental conservation and resource concentration. The prevailing ideology, she argues, is essentially anti-development and she concludes:

> It is clear . . . that planning cannot be disregarded as a definite force behind social change and that conservative anti-development and *laissez faire* policies not only protect the interests of the middle classes but actively disadvantage the least affluent. This will result in the continuing (possibly accelerating) social polarization of rural areas in the future and a reinforcement of middle-class exclusivity in the countryside. (1987: 198)

Cloke & Little (1990) go on to show that this process may well become self-reinforcing for as more middle-class incomers take up residence they will tend to become involved in local politics and will work to ensure that these planning policies remain in existence. Furthermore, they argue, the planning system itself is structured in such a way as to demand a relatively high standard of education from those who choose to participate in it. Again, the middle class will be favoured as its members have the assets

16

required to make successful representations. Thus, the middle class comes to dominate not just particular forms of housing, labour markets and communities, but also political institutions. As the processes of domination unfold they begin to accumulate; they thus reinforce one another.

In this section we have reviewed some of the most influential studies of the middle class in rural Britain. It should now be clear that the study of class has not been entirely absent from the corpus of rural studies (although the work reviewed here should not be seen as representative of the discipline as a whole) and that conversely the study of the rural can usefully inform class analysis more generally. However, there is a need to make explicit the way in which many of the processes of change in rural areas can be integrated using a conception of class formation. The various studies we have quoted here all contribute in different ways to understanding how class can be considered as the outcome of a multiple set of processes. Furthermore, it is important to note that rural space is not just a "reflection" of the middle class or merely a "stage" upon which certain middle-class actions are conducted; it is an integral part of the processes of middle-class (trans)formation.

The political activities that aim to protect middle-class spaces must be seen as part of this wider process of class formation. Individuals join groups around specific issues and act according to their perceived interests. Interests are not derived from position in some (notional) class structure but are constructed during the process of making decisions and acting upon them, that is, during the process of formation (we make this argument in more detail in Marsden et al. 1993: 138–9, see also Hindess 1987). This process always takes place in real situations and this focuses our attention on the social conditions in which interests are constructed and mobilized. Where individuals, using property, cultural or organizational assets, come to occupy common organizational or residential spaces then they are likely to influence one another in their formulations of their interests and the appropriate modes of action. Having made these points we can draw into our analysis the work that has been undertaken on middle-class action and environmentalism.

The politics of preservation

It is action over the environment that provides a useful illustration of how the processes of class formation may be part and parcel of activities which ostensibly have no class complexion. In this section we briefly sketch out how environmentalism has been considered in class terms.

In general, the rise of environmental (and other) pressure groups has been characterized as the birth of "new social movements" and it has been asked many times whether they signal the decline of (traditional) class politics and the emergence of radically new political concerns. For Touraine (1988: 68) "the notion of social movement is inseparable from class". The key distinction, in his view, lies between *social* class consciousness – that is "a social movement that is always present, even if in diffuse ways, as soon as there is conflict over the social appropriation of key cultural resources" – and a *political* class consciousness "which ensures the translation of a social movement into political action"(69), political action which is oriented primarily towards the state.

Cotgrove & Duff (1980), on the other hand, ask whether the growth in environmental protest is indicative of a fundamental change in social values. They see environmentalists as holding a set of beliefs which question fundamentally many of the core assumptions upon which industrial society is based. They attribute these beliefs to the main constituency, for "environmentalists are drawn predominantly from a specific fraction of the middle class where interests and values diverge markedly from other groups in industrial societies" (340). This specific fraction is mainly made up of public sector professionals and what this group brings into the political system are "demands stemming from non-economic values" (344). Eckersley (1989) also endorses this social profile of the most effective new environmental groups. However, she quotes Offe to the effect that the new social movements are not exclusively composed of these "new" middle-class radicals, there are two other identifiable strata that tend to form a loose political alliance with this class fraction: these are "peripheral" or "decommodified" groups who stand outside the formal labour market and hold many of the same views of industrial society as the "new" middle class and elements of the "old" middle class such as shopkeepers, farmers and artisans whose economic interests may occasionally converge with those of the "new" (as in the case study of Newby et al. 1978). Eckersley concludes that the "new" middle class, by virtue of its high education and relative autonomy from the production process will be in the vanguard of participation.

New social movements cannot therefore be solely equated with the middle class, although social movements in general are inextricably bound up with class. However, we wish to turn this problem around and ask what rôle do those political actors that have often been described as "new" and "social" play in processes of class formation? In particular, how do groups concerned with the environment and conservation tie into the spatial formation of the middle class?

Buller (1984) usefully shows how citizen action over environmental issues in the United Kingdom goes back to the nineteenth century. Many of the largest environmental groups in existence today were formed in response to fears over the "national heritage", in which the countryside has always played a prominent rôle. The Commons Preservation Society, the National Trust, and the National Society for the Protection of Birds were all formed in the latter part of the century and in the early years of the twentieth century they were bolstered by the Council for the Preservation of Rural England, the Ramblers' Association and the Youth Hostels Association. One important consequence of these groups, noted by Lowe (1977), is that they provided a stimulus to the formation of local groups by demonstrating how opposition to unwanted development could be successful and, by themselves, operating at a local level.

The extent of this type of activity has recently been investigated by Parry et al. (1992) in a sample survey of 1,600 people across England, Scotland and Wales as well as a further 1,600 men and women and 300 leaders in six contrasting localities. They found that "only a minority of the population . . . could be termed political" (8) ("political" is defined as "action by citizens which is aimed at influencing [public] decisions" – 16). However, the wealthiest quarter and particularly the top 5% participate at high levels while the poorest quarter are well below average levels of participation. The most important issue for those questioned was "environment and planning" (15.3%) and this also tended to be top of the agenda for more categories of respondent than any other. Moreover, it became of increasing importance the further up the salary scale the respondent, and leaders of action/conservation groups were on the whole "middle-aged, middle class and male" (350). Furthermore, nearly a fifth of activists had degree level education compared to 4.3% of the inactive.

These accounts place the changing social composition of the countryside in the wider context of changing forms of political activity, particularly amongst those social groups who are now beginning to predominate in certain rural areas (such as in the South East); they are likely, as Eckersley points out, to be in the "vanguard" of participation. Furthermore, these activities, as the Berkshire studies demonstrate (Barlow & Savage 1986), may well reinforce, as well as derive from, the spatial clustering of these middle-class actors. In a study of environmental and amenity groups Lowe (1977: 42) argues that

> in various ways, the results of political decisions on the environment, and therefore the activities of environmental groups that seek to influence these results, may have distributional consequences. Furthermore, organized public pressure on environmental matters has an uneven geographical spread and a restricted social base. In addition its consequences are superimposed on a scene in which amenity and accessibility are already unequally distributed.

While he concurs with the view that only a small minority of the population participate in these groups, Lowe goes on to show that the formal political structures through which these groups attempt to gain recognition for themselves may privilege access for some over others. Middle-class association and action does not take place in a vacuum. It is usually focused on precise issues and must be adapted to the formal rules of participation. Well resourced groups (in terms of cultural and economic capital) will clearly be more effective in their representational activities than groups which lack resources. Lowe notes this and points to the way that such resources can facilitate flexible strategies in the process of representation. Thus, groups may seek to influence political processes at different levels. In rural areas certain formal decision-making structures become central to the groups' activities, most notably planning. Thus: "the organization of the process of decision-making in planning may be crucial in determining the effectiveness of different types of group representation to the extent that decisions made at one stage, scale or level may foreclose options at another" (47). Furthermore, "the [planning] authority may choose to approach only those groups with which it already has a working relationship . . . those groups with which the planners feel they can have a meaningful dialogue" (48). Lowe concludes that the activities of the groups will tend to accentuate the advantages enjoyed by the well-off.

It is clear then that a crucial component of middle-class formation is the extent to which individuals enter into associations with others in order to establish or protect patterns of social and spatial privilege. As we have seen, Savage et al. (1992) suggest we might usefully consider these patterns in terms of assets, particularly associated with culture and property. They argue that cultural assets are increasingly becoming tied into petty property assets, notably owner-occupied housing. But it is not simply the house which incorporates these assets, it is also to the neighbourhood and the locale. Rural areas, particularly villages, seem to be favoured locations for a certain sort of middle-class lifestyle. Once individuals have achieved their "place in the country" they are quite likely to use diverse political means to protect this locale and the way of life it represents.

The rôle of the state: planning and development

The rise in localistic environmental protest has also been closely related to the continual development of national and local legislation associated with conservation. Lowe (1977: 38) believes that it is the decisions and actions of local authorities that the majority of middle-class groups are aiming to influence. More specifically, he argues, local groups appear to have arisen in response to the planning system and the rapid rise in the number of groups coincides with official encouragement for public participation in planning (around the time of the Skeffington Report, which placed particular emphasis on the importance of group representations, in the late 1960s). The existence of the planning system, therefore, provides an arena for middle-class political representation. Moreover, within this arena decisions are taken which have real material effects. Planning thus provides not only a focus for middle-class representation but a means whereby these representations can be translated into spatial forms. And, as Little (1987) warns, the exclusive nature of this space is likely to be accelerated as middle-class action ensures that planning enshrines these anti-development objectives.

Given its importance in the processes of social and spatial change in the countryside, and as an arena for the politics of (class) representation, we wish in the rest of this section to outline how the planning system has come to play this rôle. We show that right from the start it enshrined certain assumptions about its functioning which have since been exploited by key social groups, not least the middle classes.

The two great post-war legislative acts which determined the course of rural Britain are, of course, the 1947 Agriculture Act and the 1947 Town and Country Planning Act. It seems evident that these two acts were "wrapped around one another". Coming out of a set of inter-war and wartime concerns, and shaped by the pronouncements of the Barlow and Scott Committees, agricultural policy, with its overriding concern of increased food production, sought to establish the predominance of agriculture in the countryside. Planning policy, with its overriding concern for urban containment, also upheld the notion that rural space was agricultural space. Planning sought to protect agricultural land from external threat and urban encroachment. Furthermore, as Newby (1980a,b) has emphasized, neither agricultural policy nor planning were particularly concerned about the social implications for the countryside: agricultural policy because of the experience of depression and food shortage; and planning because it sought to preserve the countryside and its inhabitants in the "fallacious belief that the 'traditional rural way of life' was beneficial to *all* rural inhabitants" (Newby 1980a: 266; emphasis in the original). In the rest of this section we look briefly at the planning side of the rural "fence", charting its evolution during the post-war period. This account will help place the land-development processes investigated in later chapters in a broad planning context.

In their review of the first twenty years of the British planning system, Hall et al. (1973: 39) show that it came to embody three important interrelated objectives: urban containment; protection of the countryside and natural resources; and the creation of self-contained and balanced communities. Two other subsidiary objectives are also noted: prevention of scattered development and the building up of strong service centres. At the local level they see these broad objectives as crystallizing into two dominant concerns: enhancement of accessibility to services and the maintenance of a physical and social environment of quality. Although these objectives are largely *physical* in character, Hall et al. discern an underlying agenda of social concerns which they trace back to the origins of planning philosophy at the turn of the century. The key contributor was Ebenezer Howard who promoted, in his book *Garden Cities of Tomorrow* (published in 1901), the single objective of simultaneously maximising accessibility and environmental quality. Broadly speaking, Howard equated the urban with accessibility and the countryside with environmental quality. He wished to maximize the benefits of each in a new settlement form – the "social city" – which would enable residents to gain the benefits of modern societies (jobs, amenities, services) while retaining what he saw as a vital connection to nature. He thus advocated urban containment and the development of the "garden city".

The principle of urban containment was reinforced, Hall et al. argue, by Abercrombie ("certainly the most influential professional planner in the 1930s and 1940s" – Hall et al. 1973: 45) in his book *Town and country planning* published in 1933. However the case for containment was now made more explicitly on *aesthetic* grounds:

> The essence of the aesthetic of town and country planning consists in the frank recognition of these two elements, *town* and *country*, as representing opposite but complementary poles of influence . . . With these two opposites constantly in view, a great deal of confused thinking and acting is washed away: the town should indeed be frankly artificial, urban; the country natural, rural. (Abercrombie 1933: 18–19, quoted in Hall et al. 1973: 45; emphasis in the original)

Abercrombie's aesthetic concerns were also expressed through the Council for the Preservation of Rural England (CPRE), which he and others established in 1925. The main objective of the CPRE was to protect rural land from urban sprawl and from the moment of its inception "it immediately began to wage a ceaseless war . . . against the invasion of the countryside by speculative building . . ." (Hall et al. 1973: 49).

The notion of "urban containment", which had been pivotal to Howard's ambitious solution to many of the most pressing social problems, was now being employed simply to protect the "rural". The other major influence which Hall et al. identify as integral to shaping this view of planning was the Land Utilization Survey of Britain, headed by the geographer L. Dudley Stamp. Under his direction the survey had revealed that, in the years 1927–39, there had been an average annual loss of 25,000 hectares (ha) of open land to urban and industrial development. Stamp argued the case for the protection of agricultural land and "became one of the most tireless and outspoken critics of urban sprawl" (49).

The argument for "urban containment" became, therefore, an argument about the merits of protecting agricultural land as a priceless national asset. The perceived value of this asset was considerably enhanced by the wartime and post-war requirement to boost food production in the face of acute shortages and rationing. This again seemed to require a policy of physical controls over land use. However, Hall et al. discern an underlying social agenda concerned with "the preservation of a traditional way of life and a traditional economy *whatever the cost*" (51, emphasis in the original) with a subsidiary argument associated with the aesthetic beauty of rural areas. Hall et al. (1973: 52) conclude, all this "was to give the new planning system a pronounced preservationist bias". Right from the start the planning system embodied certain values associated with the uses to which rural space "ought" to be put.

In 1947 the Town and Country Planning Act established a universal compulsory planning system. A government ministry was given overall responsibility for planning matters but the actual implementation of planning decisions was left to local government, with procedures standardized from above. The administrative counties and county boroughs were the planning authorities and they were immediately required to appoint qualified planning staff to fulfil two tasks: (a) to prepare a Development Plan for land use in the local authority area for the next twenty years, reviewed every five years (and to be approved by the Ministry, thus standardizing all the various documents); (b) to control land development in the light of the approved plan ("development control work"). All proposals for development were to be submitted to the planning authority and were to be judged in accordance with the plan. An appeal mechanism was established whereby developers could object to planning authority refusals. The Ministry would establish a team of planning inspectors and these would be assigned to hear the appeals and then advise the Minister on the appropriate decision. Thus, ultimate control lay with the Minister.

As the system evolved its components became more sharply defined. Development itself was considered to be "the carrying out of building, engineering, mining and other operations in, on, over or under the land, or the making of any material change in the use of any building or other land" (1971 Town and Country Planning Act S.22). Exceptions to this were provided by a General Development Order and changes of use were covered by a definition of uses known as a Use Classes Order. The fundamental features of the system identified by Reade (1987: 11–13) were the "vague" nature of the defined uses which, he argues, relate to "nominal" uses rather than "socioeconomic realities", and the exclusion of agriculture altogether from planning control.

Hall et al. (1973: 85) note that the actual interpretation and implementation of planning procedures locally was strongly influenced by local conceptions of planning's "proper rôle". This was partly as a result of the political dimension which derived from the involvement of locally elected councillors on the planning committees, responsible for local decision-making. Ostensibly they were to act on the advice of planning officers but in reality practices evolved whereby the officers came to understand the concerns of the committee members and the members came to understand the formal rules and policies from the officers. Hall et al. (1973: 86–7) note that in the counties "the leading councillors are mainly concerned with preserving the countryside and the attractive towns and villages from urban development" while the county planners "often share the strong preservationist and conservationist sentiments of their county councillors". However, the latter were also aware of the need to accommodate *some* development *somewhere*. For it soon became clear in the early post-war period that many of the assumptions which had lain behind the new system were simply misguided. The most important was the growth in population; the planning pioneers had conceived of a balanced population, primarily between the urban and the rural, easily contained within the system's organizational framework. But the post-war years saw a rapid increase in population and this was bolstered by increased income levels in the late 1950s, manifested in the desire of many middle class people for a suburban or rural home. Thus, development pressure was more ubiquitous than had been anticipated.

The county planners had, therefore, first to convince their elected members that part of their remit was to think more "strategically" about development issues and, secondly, to find mechanisms for coping with the pressures. However, the solutions were tinged with the same preservationist attitudes. Hall et al. (1973: 87) argue that the dilemma was resolved by "designating certain towns and villages to receive new development often . . . by the negative criterion of those places which are considered not worth saving from new development because of their unattractiveness. In this way the county planners tried to provide new homes in the counties while preserving the rural amenities of the countryside". Rural villages and towns were differentiated between those that were to be "saved" from new development and those that could be "let go" (known in planning terms as "growth centres"). Places where development would be permitted were therefore designated and the types of development to be approved were specified, all within the context of development plans. Thus, the concept of a hierarchical settlement pattern containing "natural" service centres, serving hinterland rural areas, became instilled in British rural settlement planning. Policies were adopted which promoted the centralization of services and population growth while at the same time allowing for the protection of smaller outlying villages. Thus, the development trajectories of rural places became enshrined within development plans.

The "power of the plan" also focused attention on who participated in the plan-making process. It was noted that many rural residents simply did not involve themselves with the planning system and planners took steps from the late 1960s onwards to foster the inclusion of rural communities. However, as various commentators made clear (Simmie 1971, 1974, Blowers 1980, Cloke 1983), less affluent and deprived groups in the countryside were excluded from the processes of representation about development and they thus received less and less benefit from the planning process.

In their assessment of the first twenty years of the planning system Hall et al. (1973: Ch. 12) try to distinguish the main effects – who gained and who lost – from the operation of the system. The first major consequence was unsurprisingly "containment"; the use of "green belts" to prevent urban expansion and the channelling of development into certain "growth centres" successfully protected rural land. However, the movement of population associated with this pattern of development was "socially selective" for, with the exception of new towns and peripheral council estates, it consisted almost entirely of those able to buy their own homes. The second physical consequence was suburbanization, meaning the separation of home and workplace, and the tendency for certain groups to travel (often by car) further to work. This allowed these social groups to live in more desirable locations in the suburbs or in the countryside. Hall et al. argued that the main beneficiaries of the system were the "new ruralites" and the "new suburbanites". The former were particularly well served:

> ... probably most inhabitants of the countryside – those ex-urbanites who essentially use it as a way of life rather than a way of work – have gained; by establishing a civilized British version of *apartheid*, planning has preserved their status quo. This group has probably gained more and lost less than any other; and it has been quick to seize on the implications of increased public participation in planning, so as to cement its position in planning conflicts. (406, emphasis in the original)

Here we see indications of the emergence of an "exclusive" rurality, where the mechanisms of planning are used to gain and then maintain positional status within new rural spaces. (The term "apartheid" is also interesting. Although Hall et al. use it as a description of social segregation, they do not refer to any racial or ethnic characteristics of the exclusionary practices of the new ruralites).

These outcomes were not accidental. They arose from the increasingly effective campaigning activities of particular social groups attempting to gain benefit from the planning system; a system which may have nationalized development rights but only in the context of a selective widening of private property rights amongst the growing legions of middle-class residents. Commenting upon the increase in amenity and environmental groups in the post-war period, Lowe says:

Such groups proliferate. Through their growing numbers, their developing political maturity, the effective propaganda work of national environmental groups, and the sympathetic response to their activities, particularly from official quarters and the press, they appear to exercise increasing influence over decision-making on the physical environment. (1977: 35)

The diverse sets of demands being placed on the planning system at the local level, identified in the early 1970s, had by later in the decade placed the universality of the system under pressure. The response, according to Brindley et al. (1989), was the search for new forms and styles of planning to meet the needs of different places. They characterize the early 1980s, and the efforts of the Thatcher Government to "free up" the system, as a "moment of transition" as "one dominant ideology of planning attempts to replace another" (7). This might be apparent in the urban arena (with the establishment of many urban development bodies effectively sidestepping the local planning system) but in the rural domain it is more questionable.

We have shown elsewhere (Marsden et al. 1993: Ch. 5) how Conservative neo-liberal impulses were mediated in the rural sphere. The crisis of agriculture in the early 1980s coincided with government concern to rejuvenate the rural economy through the unleashing of market forces. The Alternative Land Use and the Rural Economy (ALURE) working group composed of officials from a range of government departments was established to explore how this economy could be stimulated. An increased emphasis on a *laissez-faire* approach was evident in DoE planning circulars of the mid-1980s, particularly the draft *Development involving agricultural land* (DoE 1987) which suggested that farming should be seen as a land use only on a par with environmental and other economic uses. However, the circular was seen by conservation groups as lowering the barriers to development in the countryside and a storm of protest followed. The final version of the circular promised to "protect the countryside for its own sake", a sentiment not too far removed from those enshrined in the Scott Report and the 1947 Planning Act. The tentative moves to "free up" the planning system in the mid-1980s had, by the end of the decade, been replaced by a commitment to protecting rural land from sporadic development. The huge amount of political pressure on the government from many of its "natural" supporters in the countryside had swung the tide against deregulation (the apotheosis of this was reached with the success of the Green Party in the 1989 Euro-elections). The 1991 Planning and Compensation Act thus strengthened the rôle of planning by bringing all rural areas under the local development plan-making process for the first time. The plan was now to be paramount, with all development control decisions being "made in accordance with the plan unless material conditions indicate otherwise".

Until recently the influence of the planning system, and thus of the middle class, was only really effective in rural settlements. The bulk of rural land lay outside its remit. However, there are reasons for believing that the situation may have irrevocably changed. The main shift has been brought about through the crisis in "agriculture and agricultural policy". Overproduction, depressed world markets, huge costs to taxpayers and consumers, and the perceived environmental consequences of modern agriculture have all conspired to bring fundamental changes to what is often referred to as the "productivist regime" (Marsden et al. 1993). The changes that this "regime" wrought upon rural England are by now familiar and include increased farm size, product specialization, and the substitution of capital for labour on farms. These changes *within* agriculture also facilitated some of the other marked changes in rural England that we noted earlier. As Newby (1980a,b) shows, the restructuring of agriculture created a space in rural society which the ex-urban middle class was able to fill (notably in the housing arena, where the exodus of labour from agriculture created spare property). While the productivist regime held sway the political effects of this social change remained muted and Newby et al. (1978) could still examine how farmers retained political power in rural local government. But in many rural areas farmers were increasingly replaced in local political institutions by members of the middle class. More importantly, however, the Ministry of Agriculture (MAFF) and National Farmers Union (NFU) relationship (which has been often described as "corporatist") dominated rural policy-making from the centre. Local authorities had little control over rural land use (agriculture lay outside their remit) and any transfer of land from agriculture to other uses often required MAFF's blessing (see Marsden et al. 1993: Ch. 5). However, during the 1980s the "productivist regime" entered a crisis. Now the emphasis shifted from simple food production to diversification of the farm economy and to environmentally beneficial farming practices. As an increased number of proposals came forward for the transfer of rural land from agriculture to other uses so planning authorities found themselves more deeply involved in rural land-use issues. At the same time MAFF effectively withdrew its veto over certain types of land development and local planning authorities began to adopt development plans for rural areas.

The crisis in agriculture has thus opened up a space for new uses of rural land to be asserted. Developers have clearly seen fresh opportunities in a domain that was previously dominated by agriculture (many of these developers are also farmers) and have been encouraged by government rhetoric around "diversification" of the rural economy. However, middle-class groups, often alarmed at the aesthetic and environmental consequences of new development proposals, have become increasingly effective at voicing their opposition through the planning system, as we show below. Planners, now more centrally involved in planning rural land use, have often found themselves attempting to balance competing concerns using both forward planning and development control techniques.

These new development opportunities are not simply associated with the restructuring of agriculture, however. The increase in the numbers of middle-class residents, or would-be residents, of rural areas also opens up development opportunities. House-builders, for instance, know that there is great demand for housing in "exclusive" rural locations (i.e. where development has been restricted by the planning system) and this creates development pressure upon such places, pressure which is often resisted by existing residents. Developers may respond to the lack of opportunities in "authentic" rural locations by providing "manufactured" rural housing. The increased middle-class presence in the countryside, coupled with the crisis in agriculture, may also lead to a shift from production to consumption demands over rural land, providing farmers with further diversification opportunities. Again these may be resisted by some residents. So while we can characterize certain rural areas as increasingly dominated by a particular class, we must recognize that fissures are likely to appear between those residents who have attained "positional status" (and who will act to maintain it) and those who are "on the outside trying to get in". This all adds up to more or less development opportunities, depending on the local balance of power.

How development interests and social groups compete within the planning arena must therefore be examined, in particular places at particular times. This is the task we undertake in this book. However, as we have tried to show in this section, the planning arena is not neutral; it privileges the representations of some over others, it hears some voices and not others, and in that regard must be seen as an active force structuring the development process.

Conclusion

We began this chapter with a discussion of class. We proposed the view that classes form themselves in particular places at particular times. Partly, this formation takes place in the work place but it also takes place elsewhere, most notably in living environments. We then reviewed recent sociological work on the middle class in rural England and found that the theme of an increasing middle-class presence has been a recurring one. In part this has arisen because agriculture has freed assets (e.g. property) which have become available for social groups with a strong desire to live in the countryside. The fulfilment of this desire has been bolstered by the

operation of the planning system which, by channelling growth into certain "key" settlements, has facilitated the exclusion of certain (poorer)
groups from many rural areas and allowed the middle class to become
increasingly dominant. The formation of middle-class rural places has
also presented developers with many opportunities, provided they are
able to negotiate the planning system. The terms of such negotiations
have recently shifted with the crisis in agriculture. Now, new uses for
rural land are actively sought out. Many of these are traditional industrial
uses, but the most profitable (and perhaps least contentious) are likely to
be those that meet the demands of dominant middle-class groups. However, the crisis in agriculture has also engendered much uncertainty
about the proper uses to which rural space should be put. In this period
of uncertainty conflicts around development processes have become
increasingly bitter and have placed some amount of strain on the planning system. In all the studies we document in this book planners are constantly called upon to adjudicate between competing claims on rural
land. The outcome of these various conflicts and negotiations is the
(material and social) shape of the rural locality. Through an examination
of the various land-development processes which cut across such places
we can begin to understand how the "rural" comes to be made in the likeness of those who are primarily responsible for its making.

CHAPTER 2

Regional development and social change

Introduction

Any discussion of the processes of class formation in rural areas must first place those areas in their wider context. From the previous chapter we can see that economic restructuring and shifts in social aspirations, associated most notably with owner occupation and the desire to live in rural neighbourhoods, have come to shape rural development processes in crucial ways. As we mentioned earlier we will "ground" our analysis in one rural locality where we believe the processes of middle-class formation are leading to the reconstitution of rural space. This locality is Aylesbury Vale in Buckinghamshire, situated to the northwest of London. The area is now well known to have experienced high levels of economic growth, population increase and prosperity during the 1980s. These phenomena makes it a useful "test bed" in which to examine the arguments put forward in Chapter 1. However, in order to understand the key processes shaping rural Buckinghamshire we must first put the study area in its wider context. In this chapter we outline the development of the South East region (SE) during the 1970s and 1980s, concentrating particularly upon the western counties of the outer South East (generally known as the Rest of the South East – ROSE, see Fig. 2.1). We then go on to trace the key processes of change in Buckinghamshire and Aylesbury Vale. The description of the economic, social and political context of the study area allows us to show the key pressures bearing upon rural land in the late 1980s and sets the scene for the land-development studies which follow.

Within the UK the SE has traditionally been viewed as a growth region, the "leading edge" of the national economy, although in European terms its prosperity is unexceptional (SERPLAN 1990). It has been termed the "core" region of the UK (Massey 1984, Murray 1988) in reference to the number of political and economic institutions located in the region which are deemed to have control over institutions in other areas. Patterns of development in the region are also largely conditioned by the relationship between London and the rest of the South East. This relationship is

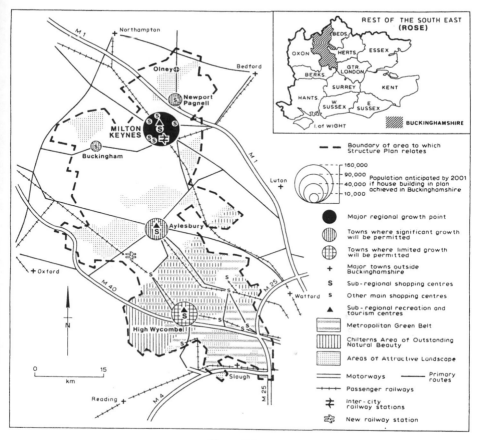

Figure 2.1

complex, with the ROSE not simply locked into a dependent relationship with the capital but also capable of generating its own independent growth areas. The structure and development of London needs only brief description. Our main concern is the prosperous subregion known as the "Western Crescent" which contains the counties of Berkshire, Buckinghamshire, Hampshire, Hertfordshire and Oxfordshire (Hall et al. 1987). This provides the immediate context for our detailed study of Aylesbury Vale. However, the development trajectory of this subregion can only be explained by reference to broader regional patterns of change. This chapter therefore considers three spatial scales – the region, the county and the district. In order to describe the context in a meaningful way, one that provides a consistent historical account, we are concerned here with two main dimensions of change – the economic and the social. Furthermore,

31

some attempt will be made to show how political/planning policies have evolved both as a response to, and as an attempt to pre-empt, these economic and social changes. This becomes clearer as we descend to the county and district levels. We are seeking here to characterize the region at the time of our study. Therefore, this chapter should be seen as a summary of the SE in the 1980s. The types of changes that interest us set the scene for our locality work which was undertaken between 1989–92.

The South East region

The SE accounts for 11% of the total land area of the UK and with its population of 17.3 million is by far the most populous region in the country (SERPLAN 1990). It accounts for 30% of the national population (*Regional trends* 1989). The population of the SE has increased in recent years and during the 1980s the region accounted for 58% of national population growth. In the ROSE area this population growth was even higher at 65% (SERPLAN 1988, *Regional trends* 1989). Within these aggregate figures there has been a changing balance of population between London and the ROSE. In the period 1961 to 1971 London lost 7% of its population and between 1971 and 1981 it lost another 9.9%. By 1981 it had nearly 2 million less people than in 1939 (Hall 1984). During the 1980s the city's population stayed relatively stable. The overall decline in London's population was more than counter-balanced by the increasing population of the ROSE. Between 1961 and 1991 the population of this area grew by over 2.5 million (*Regional trends* 1993). Projections from the Office of Population Censuses Surveys (OPCS) anticipate a continuation of population growth within the region, particularly within the ROSE area. Between 1986 and 2001 the increase is expected to be 8.9% in the ROSE and 2% in London, giving a regional figure of 6.2% (SERPLAN 1988). Furthermore, the number of households has increased by more than a million since 1970 as a consequence of the fission of the population into smaller units (SERPLAN 1990).

The patterns of population growth within the region must take account of economic, social and political factors. We shall come to the first two in due course, but in explaining the trend identified by Hall (1984), whereby the strongest population growth has moved steadily farther out from London – whereas during the early post-war years it was 15–35 miles from London, by 1961–71 it was 30–60 miles away and by 1971–91 more than 60 miles distant – we must take account of active attempts, mainly via the mechanisms of planning, to direct growth pressures within the region.

As we explained in Chapter 1, the principles of urban containment, protection of the countryside and the creation of balanced communities,

fed directly into the post-war planning system. The growth of London
was to be contained by means of restrictive green belts, "broad continuous swathes of land approximating to the rural–urban fringe zone, in
which there is a strong presumption against urban development and
within which open countryside is maintained so that recreational and
other open space deficiencies of the urban areas can be met" (Thomas
1990: 135–6). In order to protect this "green swathe" the "garden city"
idea was employed to establish a series of new towns beyond the green
belt to soak up the city's excess population (Abercrombie 1944). After
1945, therefore, with the introduction of these new spatial policies, towns
situated in a ring 20–40 miles from central London began to swell and
new towns grew out of villages on greenfield sites. By 1981 there were
11 London new towns which had a combined population of more than
800,000 (Hall 1984). These rings of growth continued to expand farther
out from the city as improved transport systems and patterns of economic
change determined new commuting patterns. Although planning policies sought to channel and control growth pressure, retaining the distinctions between town and country, growth centres and unspoilt
communities, they were not wholly successful in achieving their initial
aims. The policies were largely overtaken by events, as Warnes indicates
when he says "the transactional and communications contacts that bind
a modern metropolitan area are now a loose pattern of cascading links
with no clear break between "within-city" and "between-city" connec-
tions" (1991: 160).

Partly, this failure has resulted from the policies themselves. Thomas
(1990) argues that the green belt approach was insufficient in itself to cope
with urban pressures, while Warnes believes it should constitute "one of
a number of techniques which together implement urban policy, and its
success must depend to a considerable degree upon the effectiveness of
the other elements" (1990: 136). These other elements, Warnes argues,
were not sufficiently co-ordinated with the green belt strategy. Specifically, the new town programme did not result in diminished pressure on the
green belt area. While clearly facilitating the rapid growth of certain centres, new towns did not succeed in soaking up regional growth pressures.
Fisherman argues that the "new town methodology has become increasingly marginal to planning debate and new towns themselves increasingly marginal to the built environment" (1991: 234). He quotes Cherry
(1986) to the effect that these settlements "contributed only in a limited
way to post–war development" (Fisherman 1991: 234). "Dispersal"
proved such an irresistible force that it profoundly undermined attempts
to channel it into particular places. Furthermore, the ideal of "balanced"
communities, where individuals could live in close proximity to their
work, was upset by economic and social change, where employees operated in regional job markets and where the growth of dual–income fami-

lies made it extremely unlikely that both workers would find jobs in the same place.

The post-war planning system, applied most rigorously to the burgeoning SE region, became increasingly a victim of its own spatial logic. Its emphasis upon urban containment and rural protectionism – drawing upon strict spatial boundaries between the town and the country – became largely undermined by a regional economic restructuring process which recast the relationships between residence and workplace, communications and site location. Nevertheless, as we shall see, this does not mean that the planning system has had no effect; on the contrary, the sustained nature of these development pressures in parts of the region made the "zoning" decisions of planning authorities crucial in determining the rate of growth in particular settlements.

The regional economy

According to Murray (1988) the UK economy can be seen as divided between a southern "core" and a "periphery" which encompasses all other regions: in general, the farther from London the slower the growth. This situation derives from the location of economic control functions within the region: "the 'south' has a higher concentration of secure corporate head offices and a smaller share of vulnerable production and assembly plants than does the 'north'" (Martin 1989: 85). However, despite this apparently favoured status, all is not rosy within the "core" region itself; geographical patterns of advantage and disadvantage can be discerned. These patterns are closely related to the broader restructuring of the regional economy and have been exacerbated by the recent recession (from 1989 to 1993) which hit the SE particularly hard.

In the 1930s the new manufacturing industries grew up on greenfield sites around London. In the post-war period the emergent industries jumped over the green belt to the new and expanded towns of the RoSE. As these earlier industries declined so the newer industries located in different areas. The tendency for new sectors to be located away from the sites of earlier industrial formations meant that patterns of advantage and disadvantage were continually recreated. The most recent of these bouts of restructuring concerned the shift from manufacturing to services in the 1970s and 1980s. Between 1971 and 1984 the RoSE lost nearly 200,000 manufacturing jobs but gained 650,000 jobs in services. All manufacturing sectors, with the exception of electrical engineering, lost jobs in this period and all service industries gained them (SEEDS 1987). By 1989 17.2% of the SE's workforce was employed in manufacturing while 47.7% was employed in services (*Regional trends* 1991). Particularly important here was the growth in financial services where employment increased by 88% during the 1980s, making this sector an important engine of growth in the regional economy (Leyshon & Thrift 1993).

34

The rise and fall of the region's manufacturing industry, and the rise of the service sector, had differential effects on the subregions of the SE. This has been referred to as a "south/south divide" (SEEDS 1987) within the ROSE; between the east and west and between the industrial towns of "high Fordism" (Basildon, Luton, Bedford and Oxford) and the rising centres of the "post-Fordist" economy (Newbury, Bracknell, Basingstoke and Cambridge). While deindustrialization hit some areas particularly hard (together, Bedfordshire, Hertfordshire and Kent lost 101,000 manufacturing jobs between 1973 and 1981, 72% of the total ROSE loss – SERPLAN 1985) the new growth industries tended to avoid those areas and headed for the more prosperous locations. In the ROSE the number of jobs increased by 10% between 1971 and 1981 and by a further 2% between 1981 and 1984 (SEEDS 1987). Between 1982 and 1986 the ROSE area grew by 25%, twice as fast as the north of Britain (Murray 1988). It was noted that, as in the country as a whole, "there has been a clear inverse relation between the scale of de-industrialization and the growth of the service economy. Those areas which have suffered the highest rates of decline in manufacturing employment have benefited least from service sector development" (Martin 1989: 87). However, in general the increase in service employment more than compensated for the decline in manufacturing.

It was the growth in services, therefore, that marked the SE out most clearly from the other regions of the UK. According to the core/periphery thesis this feature of the regional economy can be attributed, at least in part, to the retention of control functions within the region. Even though the SE economy may expand into other regions "for the most part it is the routine functions which are moved out leaving the strategic functions in London and its vicinity" (Murray 1988). There is some evidence to support this view. Of the largest companies in the UK no fewer than 458 had their headquarters in London during the mid-1980s. Other places within the SE also had a sizeable number of headquarters. Berkshire, for example, was found to have 28, whereas the entire northern region (Cumbria and the North East) held only 15 (TCPA 1985). Another indicator of the SE's centrality can be found in the share of "Research and Development (R&D)" services to be found in the region. In 1976 it was estimated that the SE contained 57.4% of the UK's employees in this sector (as opposed to 9.55% in the North West for example) (Howells 1984). The major concentrations were in Hertfordshire, Oxfordshire, Berkshire and Hampshire. This was underpinned by the state's placement of contracts within the region. In 1983–4, for instance, nearly 70% of defence spending went to the two southern regions of the SE and the South West (TCPA 1985).

A major consequence of these factors was that the SE captured a disproportionate amount of the economy's lead sectors, most notably high tech industries. For instance, 38% of information technology employment in the UK was located within the SE (Hepworth et al. 1987) while evidence from the unlisted securities market (the capital market for small firms) demonstrated that the region's share of such firms in electronics and computer sectors was 72% of the total (Mason 1985).

This "leading edge" was located in very specific parts of the region with the core of the high tech activity located at the London end of the "M4 corridor" in a belt running from Hertfordshire to the northwest of London through Berkshire and into Hampshire and Surrey (Hall et al. 1987). The dominance of services was reflected through growth in other sectors during the 1970s and 1980s. Demand in the office sector was focused on precisely the same areas, along the M4 and M3/A3 corridors and on towns such as Basingstoke, Milton Keynes and High Wycombe; these were all favoured destinations for firms decentralizing from London (Jones Lang Wootten 1987, SERPLAN 1988).

A survey by Champion & Green (1992) confirmed the dominance of these western counties within the ROSE throughout the 1980s. They found that the largest relative falls in unemployment benefit claimants between 1985 and 1989 were concentrated in the SE, "especially in the 'Western Crescent' stretching from Dorset and Hampshire through Wiltshire and Berkshire into Buckinghamshire" (248). The best performing local labour markets in the survey were "highly concentrated in the South, particularly in a 'halo' around the London region and in a broad zone stretching from Dorset to the Fens" (265).

The relative economic success of the region meant that the level of Gross Domestic Product (GDP) and overall earnings remained high in relation to the rest of the UK. Between 1977 and 1989 the SE increased its contribution to total UK GDP from 33.8% to just over 40% (SERPLAN 1990). In 1987 GDP per head in the SE stood 17.9% above the UK average (*Regional trends* 1991). Levels of income in the SE also increased. In 1975, relative income per head of population was already 10% above the national average. By 1987 the figure was 13.8% (*Regional trends* 1991) and disposable income was 10% higher than the national average (ibid.). Again, these figures had the same spatial variation within the region. Within the ROSE, as ranked by per capita GDP, the best placed counties were those arranged within the "crescent", with Berkshire, Hertfordshire and Hampshire enjoying the highest per capita GDP of all the ROSE counties. Conversely, the four counties with the lowest per capita GDP – Essex, Kent and East and West Sussex – were all located to the south and east of London. Although by the early 1990s recession had hit the region particularly hard, Berkshire, Buckinghamshire, Hertfordshire and Surrey still retained the highest levels of GDP in the region (see *Regional trends* 1993).

These economic changes were inextricably bound up with changing working and living patterns. The labour force was, on the whole, becoming more middle class, well educated and female. This workforce increasingly travelled farther to get to work and it has been estimated that today's typical journeys to work are 40% longer in distance than they were 20 years ago. There has also been a marked transfer from public transport to car usage for journeys to work. Flows on the major roads in the London area were well over twice the average level for the country as a whole, while those in the RoSE were 50% higher. Car ownership itself was 13% higher than the national average in 1986, with Berkshire, Hertfordshire and Surrey displaying the highest levels of any counties in Britain (SERPLAN 1988).

The growth of the western counties also brought about a change in travel-to-work patterns. There was a marked increase in the interdependence between urban centres and their surrounding rural areas. The journey-to-work movements in the RoSE are now of an overlapping nature, with many "importing" areas also being "exporting" areas. The number of centres where the total number commuting in was over 10,000 increased appreciably during the 1980s (see SERPLAN 1986), with the result that the region could now plausibly be referred to as "multi-centred" (SERPLAN 1990: para. 1.30).

The economic changes which have swept through the SE up to the late 1980s clearly indicate new patterns of uneven development. While the post-war planning system sought to put in place a set of policies which would guide development in a rational and coherent fashion, these have been undermined by the volatilities governing economic life. However, as we shall show in much greater detail later in this book, the social changes which have accompanied economic restructuring have actually resulted in some strengthening of local planning policies. It is to a broad characterization of social change in the SE that we now turn.

The social structure of South East England

The economic restructuring of the SE finds its counterpart in social change. While the earlier "Fordist round" of manufacturing change witnessed the growth of the new and expanded towns in the outer ring and in the London suburbs, the latest round of restructuring, from manufacturing to services, paralleled a more complex set of social changes.

The latter were analyzed in two papers by Chris Hamnett (1986, 1987a). Between 1961 and 1981, Hamnett argues, census data reveals an increase in the region's professional, managerial and intermediate (PMI) category of over one million, or 47% (this refers only to economically active males). On the other hand, the skilled manual category declined by 413,050 or 25%. The semi-skilled personal service and junior non-manual category fell by a similar proportion, 23%, while the unskilled category fell by 159,520 or 40% (Hamnett 1986). The SE gained the same proportions of PMI workers and a higher proportion of self-employed than England and Wales, but lost a larger proportion of skilled, semi-skilled and unskilled workers. As a result, the higher socio-economic status of the SE in 1961 had been reinforced by 1981. The proportion of the PMI group in the SE increased by 11.9% compared to an increase of 9.9% in England and Wales (ibid.). In the ROSE, the PMI category increased from 22.2% to 35% of the total. The size of the increase in the PMI category in the ROSE was more than seven times larger than the increase in Greater London. By 1991 the number of professional and managerial workers in the ROSE stood at 39.3% and at 37.8% in Greater London (*Regional trends* 1993).

Hamnett (1987a) goes on to show how these social changes altered the region's housing market. Within the context of an overall increase in the number of households in the region of 10.6% between 1966 and 1981, the increase in the number of PMI households was split unevenly between London (16%) and the ROSE (57%). Skilled manual households decreased by 36% between 1966 and 1981, but only 8% of this decrease took place in the ROSE. Similarly the semi-skilled personal service and junior non-manual group decreased by 22% in Greater London, but by only 4% in the ROSE. The unskilled group declined by 37% in Greater London and 22% in the ROSE. Changes in housing provision are linked to these changes in social structure. The number of privately rented households declined by 750,000 or 43%, but this was offset by an increase of almost one million owner-occupiers (by 37%). By 1981 the proportion of privately rented households in the SE had fallen by half from 32% to 16%, while the proportion of owner-occupiers increased from 46% to 58%.

The results of these shifts in tenure were differentially dispersed within the region. London lost 4%, or one million, of its total number of households, mainly accounted for by the decline in the private rented sector. In the ROSE there was an increase of seven million or 24% in the number of households. This was mainly accounted for by the growth in owner occupation. This housing sector grew by 207,000, or 20%, in London, but by 742,000, or 48%, in the ROSE. The ROSE area accounted for 78% of the overall regional increase in owner-occupation. Commenting on these changes Hamnett concludes that they have tended to reinforce the segregation of the region into a rented core and an owner-occupied outer ring.

Although Hamnett is mainly concerned with the period before 1981, the trends he identifies clearly continued throughout the 1980s. This was most evident in the realm of private-sector owner-occupied housing. In 1976 in the region as a whole, private building accounted for 45% of all completions; in 1986 the figure was 83% (SERPLAN 1988). During the same period the number of houses in the private owner-occupier market increased by 10% (Murray 1988). Furthermore, despite the increase in the stock of dwellings (6% between 1981 and 1987 – by far the highest in the UK) prices increased markedly (*Regional trends* 1989). The average dwelling price for building society borrowing in the SE in 1987 was almost £60,000, one-and-a-half times the UK average. This price differential between the SE and the rest of the UK (plus the decline in the privately rented sector) led Forrest to comment that "it would appear that the owner-occupied market in the SE . . . is becoming a closed shop, an exclusive club with escalating membership costs" (1987: 1629).

While the ROSE as a whole witnessed the sharpest increase in owner-occupation during the 1980s there was a western "epicentre" of growth, with Buckinghamshire, Berkshire and Hampshire having the sharpest increases. This brought a certain number of problems in its wake. A survey of employers in Berkshire by the House Builders Federation (HBF) found that two-thirds of the sample had difficulties recruiting staff due to differentials in house prices between the county and the rest of the UK (HBF 1986).

As Forrest indicated, the owner-occupied market was becoming a more exclusive club, segmented according to socio-economic status. There was some evidence that purchasers in Berkshire, Hertfordshire, Oxfordshire and Buckinghamshire were more likely to be professional or managerial employees in well paid jobs and were more likely to be further up the housing ladder than the first time buyers concentrated in the eastern counties of the ROSE (Barlow 1988). It was this social group which was associated with the growth of the service industries in the "Western Crescent" so it comes as no surprise to find that they dominated the housing market in that area.

The growth of this stratum, and its specific location in the western sub-region, focused the attention of many commentators on the so-called "service-class" (see Abercrombie & Urry 1983, Savage et al. 1992). In the SE, as we have seen, service industries expanded rapidly, particularly in London and the western counties of the ROSE and it is the non-manual employees in these industries which have been referred to as members of the "service-class". This consists of professional, managerial and white collar workers, united by their employment in service industries (both within manufacturing industry and in the service sector itself) and state institutions (Abercrombie & Urry 1983). Some commentators have gone on to propose the constituents of a "service-class lifestyle" as, at the very least, consisting of owner-occupiers living in highly self-serviced homes (Thrift 1987a). We do not want to enter into a debate here about the relative merits of the term "service-class"; suffice to say that the shifts in the economy and social structure of the region in the past thirty years do seem to have increased the presence of the middle classes, particularly in the western counties which are of most interest to us.

For instance, Fielding (1989) examined the OPCS data for the period between 1971 and 1981 to try and establish the nature of out- and in-migration to the SE. When the figures for in-migrants and out-migrants are combined, the 1971–81 period reveals a loss from the middle class in the region of 21,000. While this might seem a surprising result, for other social classes the losses are higher; for example, the white collar working class lost 42,800 and the blue collar working class lost 45,200. This development leads Fielding to comment that "in so far as the rates of loss for the working class were higher than those for the middle class the net effect of the migrations was to slightly enhance the bourgeois character of the South East" (1989: 31). This, however, is not the whole picture. When those who were outside the labour market in 1971 are added into the equation these trends are exacerbated, with over half (52.3%) of the in-migrant entries from education going into the middle class. Fielding remarks that "the main social class effect of this part of the in-migration stream was to greatly enlarge the service-class component of the population of the South East region" (1989: 34).

This finding encouraged Fielding to develop the notion of the SE as an "escalator region" (1992). By moving to the SE employees have a better chance of gaining entry into the middle class: "a male white collar worker living in owner-occupation in 1971, who migrates to the South East had a better chance of gaining entry into the service-class (55.4%) [than] a female non-migrant blue collar worker living in council accommodation in 1971 [3%]" (7).

The notion of the SE as an "escalator" refers to younger workers who may enter the region, join the middle class and then leave the region later in life ("step off the escalator"). In 1985–6, for example, there was an influx into the region of 35,000 in the 15–30 age group. However, the region was a major net loser of middle-class members in the older age groups, losing 25,000 between 1971 and 1981 (Fielding 1992: 7–11). Fielding says: "people at or around the age of retirement tend to leave the London metropolitan city region to live in areas characterized by free-standing provincial cities; small market towns and villages, located in the southern half of the country" (1992: 9). Patterns of middle-class mobility were also summarized by Savage et al.: "Individuals are highly geographically mobile in their early working life (often associated with movement to and from institutions of higher education), and tend to move to the sorts of areas (such as Berkshire for computing workers) where their specialist jobs are in plentiful supply. Once located here they are most reluctant to move away and tend to become geographically static" (1988: 465).

It seems therefore, that from middle age onwards there is a desire to move out of the conurbations to smaller settlements, usually in more favoured areas of the SE. Clearly the middle class is very well represented in the western counties of the RoSE, such as Berkshire, Hampshire and Buckinghamshire. The Berkshire middle class became perhaps the most thoroughly researched through a number of papers published in the late 1980s (Barlow & Savage 1986, Savage et al. 1988, Fielding & Savage 1988). Berkshire was of interest because it contains a relatively large number of jobs in high tech industry – 10% of the county's total in 1986 (Barlow & Savage 1986). Most of these new jobs were for male professional workers with 55% being for professional and managerial workers while the number of semi-skilled and unskilled manual worker jobs in the service sector fell between 1971 and 1981. The numbers of employers and employees rose by twice the national rate during that period (Barlow & Savage 1986). The middle-class predilection for owner-occupation meant that they dominated the Berkshire housing market. Between 1976 and 1984, 38,000 houses were built by private builders in the county for owner-occupation, a 17% increase in the housing stock. These new middle-class homes also tended to be built in desirable locations, reflecting middle-class residential preferences (Thrift 1987a, 1989) and the growth of middle-class jobs in those areas (Hall et al. 1987).

These accounts of middle-class formation, linked to the broad shift in the region's economic structure, came to define the "character" of the SE in the 1970s and 1980s. Fielding exemplifies this when he writes that SE England "is stereotypically a service-class region; its economy and culture are stamped by the strong presence of its stable yet cosmopolitan managerial and professional middle classes. It is, perhaps, the nearest we get in Britain to a "non-place" urban realm, a spatially extended zone of formless urban development, the product of a service sector economy, of high levels of home and car ownership and of much prosperity, personal mobility and privatism" (1992: 15). This portrayal of the SE, which parallels accounts of suburban America (c.f. Schaffer 1991), draws our attention to the key consequences of economic and social change over the past two decades. However, this broad characterization, we would argue, is somewhat overdrawn. The description of the SE as an "urban realm" directs our attention again to patterns of uneven development. We have seen that the middle-class character of the region is most appropriate to the western counties and even within this area there are distinct variations in patterns of development. Moreover, the variable nature of development results, as we will show in detail in subsequent chapters, in *some* places becoming highly valued by particular social groups. Ironically, part of the of the explanation for this can be found within accounts of middle-class formation.

Members of the middle class are seemingly concerned to live within some kind of "rural idyll" (Thrift 1989) and are keen to protect this "idyll" against development (Barlow & Savage 1986, Savage et al. 1988). However, the increased concentration of activity by the house-building industry in the SE, providing middle-class housing, increased the pressure on these locations. This situation was exemplified in the mid-1980s by the struggles concerning new settlements. A group of house-building companies joined forces under the banner of Consortium Developments Ltd with the intention of providing 15 small towns in rural locations. A series of controversial proposals in the mid-1980s resulted in the consortium losing every application. The then Secretary of State for the Environment, Nicholas Ridley, professed himself "minded" to allow one such proposal, at Foxley Wood, to go through. However, the outcome of the 1989 European elections, in which 21% of the Thames Valley constituency voted for the Green Party, led to Nicholas Ridley's replacement by Chris Patten and the subsequent dismissal of the Foxley Wood proposal. Consortium Developments disbanded shortly thereafter. As we shall argue in succeeding chapters, place becomes more rather than less significant for the various factions of the middle class and they are willing to go to great lengths to defend it.

Into the 1990s, into recession

While the Government indicated a change of emphasis in the late 1980s, recognizing the growing strength of middle-class preservationism, the problems have not disappeared. The arguments continue over the likely scale of growth and development in the region in the next decade. In 1988 the forecast of the number of dwellings to be provided in the region was raised to 570,000, some 110,000 more than was agreed between SERPLAN and the DoE three years earlier (Champion & Townsend 1990: 201). Of this total figure 59% was allocated to existing built-up areas, the rest to areas peripheral to existing settlements. The problem associated with these projections was summarized in 1990 by the then Secretary of State for the Environment, Chris Patten: "How do we reconcile the need for new housing with deep felt environmental concerns? How do we provide the houses we need without destroying the places we love?" (*The Guardian*, 9 March 1990).

The pressure on the western counties was clearly becoming difficult to manage. Hampshire County Council, for example, in 1991 refused to accept its share of the new projections and was prepared to build 51,000 new homes in the period up to 2001, 15,500 less than in the Government guidelines. The County's Chief Planning Officer argued that: "The time has come to reduce the rate of development" while the Planning Committee Chairman said: "We want to spread the load. We want to deflect the pressure for development from west of London to east" (*The Guardian*, 9 March 1990). SERPLAN recognizes that these pressures are likely to be maintained in the west, envisaging growth in the three western counties of Berkshire, Buckinghamshire and Oxfordshire remaining at 11–17% during the 1990s. It now identifies the eastern subregion as being suitable for additional development, advocating "action in the East Thames Corridor" (SERPLAN 1990).

Meanwhile the economy entered a deep recession, one that particularly affected the SE. By early 1990 there were warnings that the boom in the region had come to an end. The *Financial Times* reported that "there is a

42

widespread perception that the Thames Valley is no longer a dynamically growing region – that, because of higher costs in terms of congestion and steeper rents due to shortage of land, the area has suddenly gone "ex-growth"" (16 February 1990). According to Peck & Tickell (1992: 359) the recession which followed "was initially triggered by the chronic over-heating of the South's economy . . . The consumption boom of the mid-1980s occurred primarily in the South following equity gains in the south-ern housing markets and an associated growth in the credit economy". In the early 1990s, however, house prices fell sharply, down 19% between 1989 and 1992 (*Regional trends* 1993). The reliance on credit during the 1980s in the SE also meant that when the bubble burst in 1989 the region's financial services sector, which had been so much the engine of growth, was badly damaged and forced into a profound restructuring (Leyshon and Thrift 1993). The fall in house prices also hit the housing sector par-ticularly hard, with the number of newly completed private-sector dwell-ings in the region falling by 17,000 between 1989 and 1991 (*Regional trends* 1993). Thus, unemployment began to rise by 5%, the sharpest increase of any region, between 1989 and 1992 when it stood at 9.4%. This recession, unlike the previous one 1979–83, hit the service sector and thus the SE particularly hard, signalling an abrupt end to the 1980s boom. The serv-ice-driven economy has proved itself particularly vulnerable in the wake of the downturn, with the expectation in some quarters that the region will not recover until the late 1990s (*NatWest Economic Bulletin* March 1993).

Buckinghamshire and Aylesbury Vale: development under pressure

Buckinghamshire sits in what has been described as the "golden cres-cent". The county itself covers an area 50 miles (80 km) north to south and 25 miles (40 km) east to west. In the north of the county lies the new city of Milton Keynes (MK), which is situated in its own development area, while in the centre of the county is Aylesbury Vale District where the main town is Aylesbury. To the south of the Chilterns escarpment lie the districts of South Bucks, Chiltern and Wycombe, the latter containing High Wycombe.

During the 1970s Buckinghamshire was the fastest growing county in the UK, with a population increase between 1971 and 1981 of nearly 20%. The initial growth centre was High Wycombe where, during the 1960s, the Labour controlled council sanctioned a policy of substantial public housing investment, supported by local employers concerned at the pros-pect of labour shortages. However, by the early 1970s a coalition of con-servation groups, spearheaded by the local branch of the CPRE and the

Chilterns and High Wycombe Societies, forced the adoption of a much more restrictive growth policy in the south of the county (Healey et al. 1982). Further growth was pushed into the north of the county and this approach was steadily formalized in local planning policies. The high rate of population growth continued into the 1980s, reflecting Buckingham-shire's prime location. Between 1981 and 1991 the population grew by 10% (1991 Census). The constraints on growth in High Wycombe and the extent of the Green Belt (50,000ha) and the Area of Outstanding Natural Beauty (AONB), 40,000ha of which is located in the south of the county, meant that much of this population growth took place in the north. Thus, the growth of north Buckinghamshire was particularly pronounced, with Milton Keynes and Aylesbury both developing at a fast rate during the 1980s (the population of Milton Keynes grew by 39.2% between 1981 and 1991 and that of Aylesbury Vale by 8.2% – 1991 Census). So while Buck-inghamshire could be characterized as the fastest growing county in the UK at this time, this really referred to the north of the county (in the south population numbers increased by only 2.1% between 1981 and 1991 – 1991 Census). The rate of increase in population is expected to continue, from over 600,000 in 1991 to more than 700,000 in 2001 (Bucks County Council 1991a). Despite the recent recession, this will entail sustained development pressure throughout the 1990s.

We have argued above that attempts to channel growth at the regional level have been undermined by the nature of the economic and social changes taking place. At the local level, however, the planning system has been much more effective in steering growth into particular locations. The local planning framework has embraced the division between town and country, and attempted to reproduce particular representations of these different locations within the county. Alongside this concern with the urban and the rural, and deriving partly from it, has been a concern to channel development pressure from the south of the county into the north. This policy was supported by the Green Belt and AONB designa-tions and the preservationist pressures in the south of the county.

The evolution of this overall planning framework began in the early post-war period. At this time the growth pressures on Buckinghamshire came from within its own borders. The Statutory Plan for Buckingham-shire, approved in 1954, made allowance only for limited growth to meet local needs. There was an agreement with the Greater London Council to take overspill population from London in Bletchley (a core town of the later Milton Keynes development), but aside from this there was no per-ceived pressure from other areas (Cresswell 1974). However, in the 1960s this situation changed. With upwardly revised population forecasts there was pressure for the county to take further population overspill from London. Also at this time the pressure on the Chilterns from London was perceived to be increasing. In response to both these problems the County

Council proposed the creation of a new city in the north of the county. This proposal was adopted by the Labour Government of 1964 and five years later Milton Keynes was established. Also, during this period, Aylesbury was identified as a London overspill town under the Town Development Act (1952) and the selective expansion of four northern towns – Buckingham, Winslow, Newport Pagnell and Olney – was proposed (Cresswell 1974). The policy of channelling pressure northwards was therefore established in the 1960s and it had become an explicit part of policy by the 1970s: "The slowing of growth in South Bucks in the period 1961–71 can be attributed largely to the operation of the County Council's Green Belt and the complementary policies as it has been a definite planning aim of the County Council since 1966 to restrict growth in the south of the county and promote growth in the north" (Buckinghamshire Proposed County Structure Plan, Survey Report No 3: Population and Employment 1973: 11, quoted in Cresswell 1974).

By the time of the adoption of the Structure Plan in the late 1970s this policy was firmly established. The Structure Plan stated that its broad strategy would be to "restrain development in south Bucks, including areas defined as MGB and AONB, and to channel new urban growth to selected locations in the remainder of the county, particularly the new city of Milton Keynes". Furthermore: "the plan concentrates most new urban development into a few centres . . . namely Milton Keynes, Aylesbury, Newport Pagnell, Olney and Buckingham" (Bucks County Council 1986).

Growth was to be channelled into the north of the county and into a few selected "growth centres". However, as Cresswell makes clear in his comments on the earliest formulation of this policy, "it is quite possible that most of the increased population of the expanding towns and villages in North Bucks is coming from elsewhere other than the south of the county" (1974: 8) and he cites in evidence the designation of Milton Keynes as a London overspill town. This being the case, the expansion of other towns in north Buckinghamshire would accommodate in-migrants from London and elsewhere: "It could well be that the local authority is devoting much of its resources to solving the problems of other bodies" (Cresswell 1974: 28).

The growth of north Buckinghamshire was only to take place in selected locations. As the Structure Plan made clear, agricultural land was to be safeguarded and there was a general presumption against sporadic development in the countryside. Development in the villages of north Buckinghamshire was to be restricted to that which could be contained within their boundaries, i.e. the so-called "village envelope". Large developments would not normally be permitted and housing densities should be "appropriate to the needs of those communities" (Bucks County Council 1986). The provision of employment facilities was to be in accordance with the policy of expanding existing establishments, providing local services, meeting the needs of agriculture, and providing jobs for a local labour force. No expansion for external reasons was envisaged. The rapid expansion of both Aylesbury and Milton Keynes could be directly attributed to this policy. However, while these two towns and the other identified growth centres took the brunt of development pressure the rest of the northern area was protected by a policy of urban containment and rural protectionism.

The Buckinghamshire economy

These planning policies were necessitated by Buckinghamshire's position at the heart of the "golden crescent" which ensured that economic development was intense throughout the 1970s and 1980s. Buckinghamshire was the only high growth county in the region to maintain its momentum during the 1980s, although it slowed with the recession at the end of the decade. The most pronounced growth was in distribution, professional and technical, and miscellaneous services. All these sectors had high absolute employment numbers and registered large increases. Primary sector employment declined while manufacturing employment only increased slightly. The latter was concentrated in Wycombe and Milton Keynes while service employment was spread between Aylesbury, Wycombe and Milton Keynes. The growth of these key centres can be attributed in part to their locational advantages. In the south, Wycombe has good road links to London via the M40 and the M4. In the north, Milton Keynes has close proximity to the M1 and is also on the northwestern Intercity railway line. These locational advantages, including proximity to other centres of growth in the Thames Valley, facilitated to some extent the growth of high tech industry in the county. However, by 1981 this sector was still quite small, accounting for about 5% of the workforce, compared to 10% in Berkshire and Hertfordshire.

Originally, high tech employment was concentrated in Wycombe but by the late 1970s Milton Keynes had become a popular location for high tech firms. According to the *Financial Times* "new records are set continually in Milton Keynes" (2 December 1988). By the mid-1980s the new city increased in its attractiveness for employers; net employment growth in the Milton Keynes Designated Area amounted to 8,400 jobs in 1987–8, a 51% increase over the previous year. Approximately 4,360 of the additional jobs created during 1987–8 came from firms moving to the city, while net employment growth from within the local economy created a further 4,060 jobs. Most of this growth (73%) was in the service sector. By 1988 the city claimed to support about 2,800 businesses, 1500 of which were attracted by the Development Corporation. Some 232 companies were foreign owned, 88 from the USA and 28 from Japan (the city provides a Japanese school) (*Financial Times*, 2 December 1988). Milton Keynes, the last of the planned new towns, was designed to accommodate this rate of growth within the city boundaries which embraced developable open space (by 1988 46% of the greenfield land – 3,000ha – was still awaiting development; *Financial Times*, ibid.).

The provision of a new city development at Milton Keynes, together with planned expansion of employment opportunities in a range of smaller towns (Aylesbury, Buckingham, Bicester, Leighton Buzzard, Brackley, High Wycombe) meant that during the 1980s the Buckinghamshire labour market area, and particularly north and mid-Buckinghamshire, was seen as having considerable capacity for employment growth. The amounts of land available for industrial and commercial uses was relatively high compared with neighbouring counties (SERPLAN 1989). There was a general increase in employment levels in the county, with male employment increasing by 11.3% and female employment by 12.5% (1991 Census). Again, these increases were sectorally uneven, with manufacturing continuing to decline during the latter half of the 1980s as a proportion of employment, and the construction, distribution and service sectors steadily increasing. The dynamic nature of these industries is illustrated tangentially by the number of births of new firms in the county which was higher than in any other part of the SE. The areas of growth were principally around Aylesbury, a range of smaller towns, and the New City of Milton Keynes. Between 1981 and 1987 employment grew by 12.4% in Aylesbury Vale District and by 20.4% in Milton Keynes. Unemployment was reduced, with the largest fall in Milton Keynes (41.3%), followed by Aylesbury Vale (36.9%), giving unemployment rates of 2.3% in Aylesbury Vale and 2.9% for the county as a whole in 1989.

The problem of skill shortages began to manifest itself towards the end of the decade and although unemployment in the county has since risen, standing at 6.6% for men and 3.6% for women during the depths of the recession in 1991 (1991 Census), evidence tends to suggest that skill shortages in the county as a whole may continue. The future labour market trends point towards the "information industries" with an emphasis on managerial, professional and skilled technical occupations (IER 1988). The main growth in demand for jobs in the county is expected to be in the professional and related occupations, with smaller increases in managerial, sales, personal service and craft/skilled occupational groups. Declining demand is expected for manufacturing and traditional service industry operatives and labourers. The main increases are expected in the health/welfare, literary artistic/sports and engineers/scientists categories. Moreover, the growth sectors tend to favour part-time workers, females and the self-employed, with the male labour force remaining relatively stable. Future demands for labour will therefore lie in the areas already in short supply, whereas those with low skills or who lack qualifications will be increasingly disadvantaged and vulnerable. SERPLAN esti-

47

mates suggest a growth in labour supply of 14% between 1987 and 1995 (i.e. 45,000) and growth in labour supply in the order of 10,870 for Aylesbury Vale, and 36,560 in Milton Keynes, is expected by the year 2001.

The story of the Buckinghamshire economy during the 1980s is undoubtedly a tale of success. It is not surprising that during this period it had the second highest growth in average weekly earnings in the SE. By the end of the decade Buckinghamshire had the third highest level of earnings outside London. The economy weathered the storm of the early 1980s recession, growing continuously throughout the decade. It mirrors to a great extent the broader regional shift that we noted earlier, from manufacturing to services, with construction, distribution, transport and other services accounting for almost 74% of the county's employment by the end of the decade. However, while Buckinghamshire epitomized, to a great extent, the 1980s boom, it also provides an insight into the longer term volatility of these shifts in industrial structure. With a new decade came a new economic climate, one much less favourable to the endless expansion of the service sector. As we noted above, the recession of the early 1990s hit the SE particularly severely and Buckinghamshire is no exception, indicating the precariousness of sustainable economic growth. However, while economic growth has been a catalyst for population inmigration (especially in the north of the county) further job growth to sustain this new population will continue to place considerable development pressure on the non-urban areas and make the problems of protecting rural land more acute in the following decades (Bucks County Council 1991b).

Social change in Buckinghamshire

The buoyancy of the local economy and the favoured location of Buckinghamshire in the "golden crescent" assured continued population expansion during the 1980s. A rather detailed picture of the socio-economic complexion of this expansion can be derived from local area housing market statistics, which tell us something about the kinds of people moving into the county (before the onset of recession). In the year ending March 1988, for example, 68% of Nationwide Anglia's mortgage loans were made to previous owner-occupiers, a much higher proportion than nationally. This high proportion of previous owner-occupiers contributed to (and was restricted by) the high prices in the county during the 1980s boom. The average price paid by all Nationwide Anglia borrowers in the county was £74,000, 51% higher than for the UK as a whole (£49,050). Prices ranged from £60,000 in MK to £100,000 in Chiltern District (Aylesbury Vale stood between the two at £69,000).

A much higher proportion of the Society's borrowers in Buckinghamshire were in managerial/professional occupations (40%) than nationally (32%), reflecting the growth of the relevant occupations and proximity to

London and other employment centres (e.g. the Thames Valley). The proportion of managerial borrowers was particularly high in Chiltern District where 60% of home buyers were in this occupational category (in Aylesbury Vale the figure was 36%). Milton Keynes had the highest proportion of skilled and semi-skilled manual borrowers (38%) compared with 18% in Chiltern. The average household income of Nationwide Anglia's borrowers in Buckinghamshire was 21% higher than for the Society's borrowers nationally. The housing market displayed all the characteristics of economic prosperity at this time –high prices and high income purchasers moving into favoured locations, particularly in the south of the county. The Chilterns were particularly attractive due to the success of conservationist pressures and the restraint policies determined by green belt and AONB status. The 1991 census showed that 73.7% of households in Buckinghamshire were living in owner-occupied accommodation, an increase of 11.1% over the previous decade.

This broad pattern of change, and the social attitudes which accompanied it, were the subject of an extensive household survey undertaken by the authors on behalf of Buckinghamshire County Council in 1991 (see Appendix 1). A stratified random survey of 2,000 households dispersed throughout the county – north and south, urban and rural – were interviewed in order to uncover attitudes to planning policies. We can use this survey data to draw out some of the broad changes in the social context and the attitudes of current residents to patterns of development in the area. The results in Tables 2.1 and 2.2 are given for the county as a whole and are divided into urban and rural respondents (as defined in the Bucks County Structure Plan). First we identified how many people had lived in places other than their current residence (at the time of the survey).

Table 2.1 Buckinghamshire Social Survey: lived elsewhere (%).

	Urban	Rural	Total
No movement	11.2	13.9	14.2
Same town/village	36.5	33.3	32.4
Elsewhere in Bucks	19.5	29.6	25.2
Outside Bucks	31.8	33.3	32.4

Table 2.1 shows the extent of residential mobility in Buckinghamshire and its spatial distribution. Clearly a minority of respondents had never moved, while the majority of those that did so moved within the county itself. The main distinction between the urban and rural residents was the numbers of urban residents moving within a town while many rural residents had moved from elsewhere in the county. The numbers moving in from outside the county were roughly comparable. The principal reasons for moving, given by the whole sample, were work (31.7%), housing type

(16.8%), the environment (15.1%), marriage (12.9%), housing cost (8.9%), and retirement (6.5%).

The survey of 2,000 households was categorized according to the Registrar General's classification of socio-economic groups (SEG). The make-up of the sample was as follows: SEG 1 (higher managerial/professional) 5.1%; SEG 2 (intermediate managerial) 20.7%; SEG 3 (supervisory and clerical) 21.9%; SEG 4 (skilled manual) 18.9%; SEG 5 (semi-skilled and unskilled workers) 2.4; and SEG 6 (economically inactive) 30.9%. The majority of the sample were in the top three groups and could conceivably be described as "middle class", while the last group, the economically inactive, may have a significant retired middle-class component. This can be broken down further into the urban and rural distribution of the classes:

Table 2.2 Buckinghamshire Survey: distribution of socio-economic groups (SEG) (%).

	Urban	Rural	Total
SEG1	3.4	6.6	5.1
SEG2	19.5	21.4	20.7
SEG3	22.0	21.7	21.9
SEG4	21.3	15.9	18.9
SEG5	2.5	2.0	2.4
SEG6	29.2	32.1	30.9

The results show distinct similarities between the urban and rural areas, with the rural respondents falling marginally more readily into the highest social classes and the retired, while the urban respondents were more highly represented in the middle categories. We may go on to ask whether "middle-class preferences" can be associated with groupings thought to represent this class. There is some evidence to support this: for instance 71.7% of SEG 2 and 80% of SEG 1 said the "scenery and the character" of the area in which they lived were important to them; 63.2% of SEG 1 supported the controlling of development in villages. Such figures do not however, give a very clear indication of what middle-class preferences might be and how they differ from the preferences of other classes.

We can now narrow the sample down to concentrate upon the rural constituents of the survey. These comprise seven villages, three of which are in the south of the county (Lacey Green, Stokenchurch, and Botley and Leyhill), three are in Aylesbury Vale (Steeple Claydon, Haddenham and Newton Longville) and one close to Milton Keynes (North Crawley) (see Appendix 1 for more details). These village survey areas provide a transect through the spine of the county. In planning terms they encompass the less restrictive villages around Milton Keynes and Aylesbury Vale and the highly restrictive green belt areas in southern parts of the county. The current structure plan, green belt, and AONB notations place these villages very much on a hierarchy of restrictiveness concerning housing, leisure and industrial developments. North Crawley, close to the Bedfordshire and Northamptonshire border and to the new city of Milton Keynes, is particularly vulnerable to large-scale development. SERPLAN's regional strategy (1990), for instance, points to the area where these county boundaries meet as one of potential growth over the forthcoming decades, particularly as east–west road links are enhanced. In the south of the county the environment is very different. Lacey Green and Botley and Leyhill are staunch middle-class preservationist and "positional" village spaces. We can obtain a more complete social picture by examining some of the results for the sample of villages. Again, we begin by providing a cross section of the social structure of the settlements using the Registrar General's socio-economic categories.

Table 2.3   Buckinghamshire villages: socio-economic group (SEG) (%).

SEG	North Crawley	Newton Longville	Steeple Claydon	Hadden-ham	Lacey Green	Stoken-church	Botley and Leyhill	All
1	6.5	8.7	6.8	3.1	15.7	7.1		6.8
2	17.4	18.5	14.6	14.6	27.0	21.4	11.6	17.8
3	21.7	25.0	28.2	20.8	20.2	18.4	30.5	23.5
4	10.9	17.4	19.4	18.8	4.5	17.3	20.0	15.4
5	3.3	1.1	5.8	2.1	–	1.0	7.4	2.9
6	40.2	29.3	25.2	40.6	32.6	34.7	30.5	33.3

These results largely bear out the social make-up of the rural areas of the south of England that we outlined earlier. The bulk of the economically active fall into groups 2 and 3 which correspond to most definitions of the middle class. The most notable figures relate to Lacey Green, an exclusive village within the Green Belt which has the highest proportions in the top two socio-economic groups. All the villages show small but significant minorities in the skilled and unskilled manual categories, with North Crawley and Haddenham exhibiting high levels of the economically inactive.

We then asked respondents how long they had lived at their current addresses in order to get some idea of levels of mobility. The results are given in Table 2.4.

These results reflect the degree of social change in the villages of Buckinghamshire, with the majority of households living in their homes for ten years or less. Recent housing developments in north and mid-Buckinghamshire villages are reflected in the higher proportions of respondents living in their houses for less than 6 years (e.g. North Crawley and Steeple Claydon). The southern villages have slightly higher proportions of long-term residents (i.e. over 21 years) and a higher middle-class com-

Table 2.4   Buckinghamshire villages: length of time in household (%).

	North Crawley	Newton Longville	Steeple Claydon	Hadden-ham	Lacey Green	Stoken-church	Botley and Leyhill	All
Years 1 or less	3.3	7.6	9.7	–	3.4	7.1	4.2	5.0
1-5	37.0	27.2	31.1	16.7	27.0	16.3	24.2	25.6
6-10	21.7	27.2	26.2	38.5	15.7	25.5	22.1	25.2
6-10	20.7	21.7	21.4	34.4	29.2	25.5	28.4	25.7
21-30	5.4	10.9	7.8	5.2	11.2	12.2	11.6	9.1
30+	12.0	5.4	3.9	5.2	13.5	14.3	9.5	9.1

Table 2.5   Buckinghamshire villages: lived elsewhere (%).

	North Crawley	Newton Longville	Steeple Claydon	Hadden-ham	Lacey Green	Stoken-church	Botley and Leyhill	All
No response	9.8	17.4	27.2	18.8	13.5	23.5	20.0	18.6
Same village	18.5	13.0	19.4	18.8	13.5	26.5	31.6	20.1
In Bucks	32.6	34.8	30.1	28.1	55.1	22.4	21.1	32
Outside Bucks	39.1	34.8	23.3	34.4	18	27.6	27.4	29.2

ponent. Respondents were also asked if they had lived elsewhere.

Table 2.5 gives some indication of where people have moved from to their current residence. The majority have moved into the villages from elsewhere and quite a high proportion, nearly 30%, have moved into Buckinghamshire (this is particularly marked in North Crawley). Slightly higher proportions within two of the southern villages have moved within the village. Again, the south Buckinghamshire villages show less dynamism as well as a slightly older age structure. While many of the residents in the north and mid-Buckinghamshire villages were firmly on the "social escalator", whereby career move determined location of residence, for many in the southern villages the escalator has slowed as they seek the benefits of social entrenchment in village life. For instance, although 19% of the respondents in Lacey Green were aged 55–60 only half that proportion (9.8%) were present in North Crawley (see Table 2.6).

Those respondents who had moved were asked to give the most important reason governing their choice of housing location (see Table 2.7).

Work is clearly the dominant reason governing mobility; evidence of a link between this type of social change and the buoyancy of the local economy. However, the environment scored a significant second, particularly in the southern green belt villages, emphasizing that there are other factors pulling people into the rural areas which may bear some

Table 2.6   Buckinghamshire villages: age of respondents (%).

	North Crawley	Newton Longville	Steeple Claydon	Hadden-ham	Lacey Green	Stoken-church	Botley and Leyhill
17–24	3.3	9.8	9.7	8.3	1.1	8.2	5.3
25–34	14.1	22.8	24.3	20.8	10.1	17.3	15.8
35–44	30.4	21.7	27.2	27.1	25.8	20.4	27.4
45–54	16.3	17.4	18.4	9.4	15.7	18.4	16.8
55–60	9.8	5.4	1.0	2.1	18.9	12.2	11.6
61–64	1.1	6.5	1.9	4.2	9.0	7.1	5.3
65+	15.2	10.9	12.6	20.8	15.7	13.3	15.8
75+	9.8	5.4	4.9	7.3	5.6	3.1	2.1

Table 2.7   Buckinghamshire villages: reasons for moving (%).

	North Crawley	Newton Longville	Steeple Claydon	Hadden-ham	Lacey Green	Stoken-church	Botley and Leyhill	All
Marriage	10.9	9.8	18.4	11.5	10.1	20.4	16.8	13.9
Education	–	4.3	2.9	5.2	–	1.0	2.1	2.2
Housing:								
type	14.1	17.4	13.6	21.9	15.7	10.2	15.8	13.2
cost	3.3	13.0	15.5	1.0	1.1	13.3	8.4	7.9
Travel	1.1	1.1	1.0	–	–	2.0	1.1	6.3
Work	31.5	27.2	25.2	24	27.0	30.6	27.4	27.5
Environ-								
ment	20.7	13.0	10.7	16.7	33.7	14.3	21.1	18.6
Retirement	9.8	9.8	2.9	19.8	5.6	–	3.2	7.3
Other	8.7	4.3	9.7	9.7	6.7	8.2	4.2	7.3

resemblance to the search for a "rural idyll".

High levels of mobility raise the issue of "commitment" to place. We asked two questions related to this: the first concerned attachment to the neighbourhood; the second asked how long the respondent intended to stay in the neighbourhood.

In Table 2.8 the highest level of local attachment is found in the most exclusive of the Green Belt villages with high numbers of the "very attached" in the village close to Milton Keynes (i.e. where high proportions of residents [58.7%] had lived there for less than 10 years). Overall, there seemed to be strong levels of attachment throughout the villages.

Once again a large majority intend to stay in the neighbourhood (see Table 2.9). However, this majority is slightly higher in the south than in the north reflecting perhaps a recognition that once access to these villages has been achieved it should not be given up easily. High levels of residential mobility do not, therefore, necessarily negate the significance of place; they simply lead to its reconstitution.

We then sought to discover whether our respondents, having achieved their place in the country, were prepared to countenance more development in the area. We asked a number of questions related to a variety of

Table 2.8 Buckinghamshire villages: attachment to neighbourhood (%).

	North Crawley	Newton Longville	Steeple Claydon	Hadden-ham	Lacey Green	Stoken-church	Botley and Leyhill	All
Very attached	66.3	47.8	51.5	52.1	74.2	56.1	54.7	57.5
Attached	26.1	41.3	37.9	39.6	22.5	35.7	36.8	29.0
Not attached	7.6	10.9	10.7	8.3	3.4	7.1	8.4	8

Table 2.9 Buckinghamshire villages: hoping to stay in neighbourhood (%).

	North Crawley	Newton Longville	Steeple Claydon	Hadden-ham	Lacey Green	Stoken-church	Botley and Leyhill	All
Yes	79.3	72.8	75.7	82.3	85.4	81.6	84.2	80.1
No	20.7	27.2	24.3	17.7	14.6	17.3	15.8	19.6

Table 2.10 Buckinghamshire villages: attitude to more homes (%).

	North Crawley	Newton Longville	Steeple Claydon	Hadden-ham	Lacey Green	Stoken-church	Botley and Leyhill	All
Limit growth	73.9	71.7	65.0	77.1	83.1	77.6	76.8	75
More housing	26.1	28.3	35.0	22.9	15.7	22.4	23.2	24.8

Table 2.11 Buckinghamshire villages: location of new employment in the countryside (%).

	North Crawley	Newton Longville	Steeple Claydon	Hadden-ham	Lacey Green	Stoken-church	Botley and Leyhill	All
Yes	4.3	14.1	15.5	7.3	5.6	5.1	8.4	8.6
No	95.7	85.9	84.5	92.7	94.4	94.9	91.6	91.3

different development processes: housing, employment, and primary industries.

Table 2.10 shows there is clearly a majority in favour of limiting growth, shown most starkly in the exclusive Green Belt villages, but also present throughout the survey areas. This attitude also prevails on employment location, shown in Table 2.11.

That new employment in the open countryside should be disapproved of to such an extent is perhaps not surprising. Only in Newton Longville and Steeple Claydon, where some traditional industrial development was still in existence, was there a significant minority of residents favour-

Table 2.12 Buckinghamshire villages: location of new employment in villages (%).

	North Crawley	Newton Longville	Steeple Claydon	Hadden-ham	Lacey Green	Stoken-church	Botley and Leyhill	All
Yes	20.7	29.3	31.1	37.5	18.0	21.4	10.5	24.0
No	79.3	70.7	68.9	62.5	82.0	78.6	89.5	75.9

ing new employment in the countryside.

The opposition to new employment in villages is less emphatic than in the open countryside but is still overwhelmingly against, particularly in the south of the county (see Table 2.12).

The survey therefore indicates the extent of social change in the Buckinghamshire countryside, shedding some light on the aspirations and attitudes of rural residents. These changes can be characterized as having resulted in rural villages becoming "home" to those in the higher income groups who have a predilection for the rural environment. When asked "what is the best thing about living in the area", 72.5% of respondents in the villages said "the quiet and the green". And, having gained their place in the country, these residents seem ill disposed to any further development of the rural landscape. Clearly residents in the "exclusive" villages in the south of the county, surrounded by green belt and AONB, were particularly resistant to development in both the villages and the countryside, but such attitudes were also prevalent amongst the villagers in the north.

The changing social complexion of the county is also reflected in the politics of Buckinghamshire which, outside Milton Keynes, are dominated by the Conservative Party. For instance, since 1970 the Aylesbury Vale seat has been held by the Conservative MP Timothy Raison, and the Buckingham seat by the Conservative William Benyon and latterly since 1983 by George Walden. Table 2.13 indicates a consolidation of Conservative strength in the county with the Labour Party being replaced as the main party of opposition by the Liberals (LDP). The same trend is evident in elections to the County Council, as shown in Table 2.14.

Again, the Conservative dominance of the county's politics is emphasized. Perhaps the most notable feature of this latter table is the long-term decline in the number of independents. In rural areas independent councillors generally represent local landowners, farmers and small businesses (see Grant 1977). Their decline can be ascribed to the influx of incomers to the area who, as the figures in Table 2.14 suggest, in the main appear to be Conservative voters.

The overwhelming dominance of the Conservatives has resulted in some measure of stability in the political leadership of the County Council (in the 1993 "shire" elections Buckinghamshire was the only county to

Table 2.13 Buckinghamshire voting patterns (%).

	Year	Vote	Conservative	Labour	Liberal
Aylesbury	1970	75.6	53.3	35.0	11.7
	1974	81.7	47.0	26.4	26.6
	1974	74.4	46.8	29.0	24.2
	1979	78.0	58.2	24.2	17.6
	1983	71.5	58.1	12.2	29.4
	1987	74.5	57.5	13.9	28.6
Buckingham	1970	81.8	47.5	43.2	9.3
	1974	85.3	40.7	36.0	23.3
	1974	79.7	42.2	37.6	20.2
	1979	78.6	51.3	34.1	13.6
	1983	77.1	56.9	15.0	28.1
	1987	78.3	58.6	16.5	24.9

Table 2.14 Buckinghamshire County Council seats by party (%).

Year	Ind	Con	Lab	Lib	Total seats
1973	24.3	45.7	24.3	5.7	70
1976	20	51.4	22.9	5.7	70
1977	5.7	91.4	1.4	0	70
1978	4.3	92.9	1.4	0	70
1981	7.1	67.1	21.1	2.9	70
1982	7.1	65.7	21.4	2.9	70
1983	7.1	67.1	21.4	2.9	70
1984	7.1	67.1	20	4.3	70
1985	1.4	69	17	12.7	71
1986	1.4	69	17	12.7	71
1987	1.4	67.6	17	12.7	71
1988	1.4	67.6	17	12.7	71
1989	2.8	70.4	15.5	11.3	71

remain under Conservative Party control). This has allowed the construction of a coherent set of policies and the establishment of firm relationships between the politicians and those whose job it is to serve them, that is council officers and professional planners. The result has been the formulation of a set of policies which have persistently sought to channel growth northwards and which, in the north, have been orientated towards conserving the countryside. Thus, the urban areas of the north have been developed rapidly during the 1970s and 1980s, leading to acute pressure on public space and housing densities. This is evident when we turn to examine Aylesbury Vale

Aylesbury Vale: town meets country
Aylesbury Vale District encompasses approximately half the county of Buckinghamshire. The District lies broadly north of the Chiltern escarp-

ment, from the town of Wendover in the south to Buckingham in the north. Aylesbury is the administrative centre of the county and contains a population of approximately 50,000, one third of that of the District. As well as Aylesbury there are over 100 smaller settlements in the Vale. In the south of the District development pressure is at its most acute, particularly around the town of Aylesbury which lies only a few miles north of the Green Belt and AONB. Farther to the north however, the District still retains a traditional rural character. The policy of urban containment has been relatively successful. Although there has been a considerable amount of social change, the area still "looks" rural. According to a local amenity society: "For many people, the Vale represents the first "proper" countryside they meet when driving up out of London, less than 40 miles from its edge. The fact that it is "proper" countryside is probably its most treasured aspect to its inhabitants, and the thing they most wish to cherish and protect" (Friends of Aylesbury Vale 1986).

The pace of economic and social change in the District has been rapid. The population of the District increased by 10% between 1981 and 1989 (*Regional trends* 1991) and a further 11.6% increase is expected by the year 2001 (Buckinghamshire County Council 1991). The area has thus been on the front line of development pressure. The restructuring of the local economy meant that by 1989 Aylesbury Vale had the highest proportion of its employees working in the service sector of any Buckinghamshire district; 51.3% worked in "other services", 26% in construction, distribution and transport, and 20.1% in manufacturing (*Regional trends* 1991). Firms such as Equitable Life spearheaded the service sector shift by moving their administrative operations to Aylesbury in the early 1970s, followed by Target Life Assurance and the regional offices of Lloyds Bank. These were high profile firms within the town and represented the shift of the local economy more fully towards services. In the manufacturing sector firms such as CBS, Cadbury–Schweppes, Nestlé and the London Brick Company dominated the industrial landscape.

As in the county as a whole, Aylesbury Vale District has traditionally been dominated by the Conservatives. As mentioned above, the two MPs who sit in the District are Tories and the District Council has, for the past few years, sustained a Conservative majority (see Table 2.15):

Table 2.15   Aylesbury Vale District Council seats by party (%).

Year	Ind	Con	Lab	Lib	Total seats
1973	59.3	18.5	22.2	0	54
1976	10.3	51.7	15.5	3.4	58
1979	20.7	62.1	13.8	1.7	58
1983	0	58.6	6.9	8.6	58
1987	13.8	58.6	3.4	24.1	58

The most notable feature of Table 2.15 is the decline in the number of independent councillors on a more pronounced scale than at the county level. This allowed the Conservatives to reach a position of dominance on the District Council which lasted from 1976 to 1991. In the District Council elections of that year the Liberal Democrats gained 11 seats while the Conservatives lost 8. This led to a "hung" council with the Conservatives holding 29 seats, the Liberal Democrats with 22 seats, Labour 1 and Independents 6 (*Bucks Herald*, 9 May 1991). The Tories retained control but only through the Chairman's casting vote. The political stability of the Council, until the last round of local elections, allowed the development of a professionalized planning framework to deal with the growth pressures in the District. As we have seen, Structure Plan policies entailed the development of Aylesbury, but restraint in the surrounding countryside.

The expansion of Aylesbury dates back to 1961, and the adoption of the Aylesbury Town Map, which provided for an increase in population from 23,000 in 1957 to 42,000 in 1974. By 1981 the population of the town had increased to 50,000 while the number of jobs had increased from 16,700 to around 34,000 (Aylesbury Vale District Council [AVDC] 1989a). The rapid development of Aylesbury led to a shortfall in the amount of open space in the town and increasing levels of congestion, a point explicitly recognized in the 1989 Aylesbury Local Plan (AVDC 1989a). However, restrictions on development elsewhere in the District, as laid down in the Structure Plan, ensured that Aylesbury itself continued to soak up a major part of the pressure.

It was estimated in the Structure Plan that, between 1986 and 2001, 6,200 dwellings would be built in Aylesbury with a further 5,000 in the remainder of the District (AVDC 1991a). According to the Aylesbury Local Plan "the growth of Aylesbury has been largely 'employment-led'; that is, the migration of population to the town has been dependent to a significant extent on the availability of jobs". Therefore, to meet the expected increase in population around 6,000 new jobs would be required in the town by 1996: "split according to existing employment structure, this represents 1.8 thousand jobs in manufacturing industry and 4.2 thousand in services" (AVDC 1989a).

In order to control development in the rest of the District and meet Government guidance for district-wide plans, the Council prepared a local plan for the rural areas of the District in 1990. The general aims of the *Rural Areas Local Plan*, while recognizing the pressures for the development of land, included the following: "(a) to protect and enhance the general environment and natural beauty of the countryside, particularly those areas which are outstanding or especially attractive; (b) to protect the identity and enhance the environment of towns, villages and hamlets; (c) to protect buildings of special interest and archaeological remains; (d) to enable development to accommodate the resident and working population of the District, at home, work and play, in the right place at the right time in such a way as promotes the highest quality of life; (e) to promote a buoyant local economy" (AVDC 1991a). The emphasis here is clearly on "protection" and the "quality of life".

The Plan reiterated that "it has been a long-standing planning objective to protect the general appearance and open character of the countryside. . . . It is Government policy to protect the countryside for its own sake. This is a view wholeheartedly shared by the Council" (AVDC 1991a). While it was recognized that agriculture and the rural economy more generally were going through a period of change "the need to protect the countryside for its own sake will be the paramount consideration when assessing proposals for development" (ibid.).

The tone of this plan, and the general aims that it seeks to achieve, will ensure that development pressure is maintained on the urban areas, while leaving the countryside preserved "for its own sake". And this policy will be supported by residents in the District, as our social survey shows. The "dispersion" of development that we identified in the regional context reaches its limits at the local level; it will not be dispersed throughout Aylesbury Vale but will be firmly steered into certain growth centres which will continue to bear the brunt of economic and social change.

In conclusion, a note of caution. Aylesbury Vale did not escape the downturn in the region in 1989 and by early 1991 redundancies were becoming more prevalent with, for instance, Pitstone Cement Works closing, meaning the loss of 300 jobs, and Pitstone Brick Works following with the loss of 336 jobs. The policies drawn up in the 1980s were adopted in the context of economic prosperity and low unemployment. With a prolonged recession and increasing unemployment a different emphasis may be placed on planning policy, much more geared towards job creation as opposed to straightforward constraint. But it is far from clear that there will be any perceived need to create employment in the countryside. Quite the opposite; our survey data shows that there is desire to keep employment *out* of the countryside. However, there seems little doubt that once the recession has eased development pressure will again bear down upon the county as a whole, and the District of Aylesbury Vale in particular. One reason for this will be transport links. The A418 road which runs from east to west through the south of the District is scheduled for upgrading into dual carriageway and will form part of a link between the M40 in the west and the east coast ports. However, in the longer term it might form the basis for the M250 – a London outer orbital road which would go south from the M40 to the M4 and M3 (*Bucks Herald*, 15 November 1990). This may well increase development pressure in the south of the District. And if the favoured western subregion, of which Buckinghamshire forms a part, continues to soak up the bulk of the region's growth then Aylesbury Vale will continue to be a target for development.

Conclusion

In this chapter we have examined the uneven patterns of economic and social change in the SE. We identified the growth of manufacturing industry in the region and then the shift from manufacturing to services. The restructuring of the regional economy entailed spatial changes, leading in particular to sustained growth in the west of the region. This was paralleled by social change and the growth of the middle class.

Attempts to steer these economic and social changes have been undermined largely because the forces behind "dispersion" – disenchantment with the environment of the city and the locational advantages of the western subregion – were far stronger than the static instruments (green belts and new towns) of direction. In the absence of any semblance of regional policy (an early victim of the Thatcher Government) the forces of (selective) dispersion were allowed to prevail. The Government stated that "it is not our policy to discourage development and economic growth in the South East in the hope that it will transfer to other areas, for in that way we risk losing it altogether" (DoE 1986).

While the growth pressures on the western subregion followed the momentum of economic and social restructuring, at the local level the planning system was much more effective in directing them to particular "growth centres". This led to burgeoning growth in places such as Milton Keynes and Aylesbury, but largely preserved the countryside from intrusive development. The result of this policy was a sharper physical and social contrast between town and country (the former at least being one of the original aims of the planning system). In a county under considerable pressure, the dichotomy between the maintenance of the exclusive countryside, through its protection "for its own sake", and increasing urban densities becomes increasingly critical. Planning policies may maintain the exclusivity of rural areas for the benefit of some residents, and as our survey showed, these residents are likely to be of middle-class complexion, but the increasing cost of such policies will fall upon urban areas. Thus, the preservation of rural space comes to bestow advantages upon a few "favoured" residents while the bulk of the population suffer increased levels of development and a diminishing "quality of life". It is to the activities of the "favoured few" that we now turn.

CHAPTER 3

Constructing exclusivity

Introduction

In Chapter 1 we argued that the middle class is becoming increasingly dominant in many rural localities, while in Chapter 2 we showed the extent to which this phenomenon can be linked to shifts in the economic and social structure of particular regions, most notably the South East. We now turn to look in detail at the means by which middle-class "colonization" is achieved. In this chapter we consider the rôle of the housing development process, and the types of housing typically provided in rural areas, in fostering the process of class formation. This has two main dimensions which will be elaborated on below: first, the forms of regulation which "surround" the housing market and which attempt to ensure the delivery of particular housing "requirements"; and secondly the "market" itself, or how developers seek out and then attempt to meet or create opportunities.

It is self-evident that the middle class can move into rural areas only if housing is available. Housing becomes available if other residents leave these areas or if new housing is built. Clearly, many traditional rural residents have left to live elsewhere and much traditional rural housing has been made available to new residents. However, the amount of new stock has been restricted as housing has traditionally been subject to a variety of planning controls and regulations. Until the mid-1980s it could perhaps be argued that the control of housing development was the *raison-d'être* of rural planning and policy. Indeed, some commentators believe that the narrow and rather negative perspectives and instruments of rural planning tended to treat rural settlements very much as isolated pieces of suburbia – giving precedence to urban containment and the protection of rural land (Savage 1987). In the changed circumstances of the 1980s, however, the assumptions lying behind both housing policy and the planning system were to be seriously scrutinized. As many authors have pointed out (e.g. Saunders 1990, Ball 1986), the housing sector was to represent a key target for deregulatory reform by the Thatcher administration. New markets were to be created through the "right to buy" opportunities in public housing; developers were to be encouraged through a battery of new fiscal and macroeconomic measures, increasing their links with the

61

finance sectors; and consumers were to be given further incentives to exploit "credit-based purchasing" of a "home of their own". A detailed description of these policy changes will not be provided here, although we should note that they threw many of the traditional concerns of planning policy into doubt during the 1980s.

With the primacy given to agricultural land uses on the wane, and with considerable pressure being placed on the planning system to improve its efficiency (Healey et al. 1985, 1988), the basic post-war protectionist premises were scrutinized in the early part of the decade (Reade 1987, Evans 1991). By the mid-1980s, however, many of the direct political and ideological attacks on green belts and local planning had begun to lose their venom (see Ch. 5), signalling a significant new phase in the regulation of rural housing development. For instance, planners were now expected to produce clearer land budgets and plans concerning land availability for housing, and to do so in consultation with representatives from the housing construction sector such as the House Builders' Federation (HBF). Although these sets of revised relations have led some to identify a new form of housing *corporatism* between house-builders, central government and planners (Lowe 1988, Schucksmith 1990), this fails to capture fully the changing nature and diversity of local house-building and development, and the different local and strategic operating *contexts* in which these actors and agencies find themselves. These contexts require definition at different scales of analysis.

If the starting point for our investigation of housing development is the local level, then we can identify several key forms of land development in the 1980s worthy of in-depth analysis. First, as we noted in Chapter 2, the corporate house-building lobby organized themselves to promote the concept of the "new settlement" (Ambrose 1986, Herrington 1984). DOE figures, published in 1985, suggested an increase of 11% in households, 21% in East Anglia and 14% in the South East (SERPLAN 1986). Such statistics and calculations promoted the idea that there was insufficient land available for "brownfield development" (partly because of the over-restrictive planning system). Borrowing many of the design and social tenets first adopted in the *public* New Towns movement, corporate house-builders began to establish proposals for self-standing new "villages" or "settlements". Many of these proposals became symbolically significant as they were well publicized and highly contentious responses to housing pressures in the South East. They also put pressure on central government to define a strategy for housing developments in rural areas. (Boucher 1993). Secondly, the partial and uneven withdrawal of agriculture from villages provided the basis for a sharp growth in *farm and barn conversions*, mainly for housing (see Kneale et al. 1991). These were particularly popular in southeast and southwest England, and they were encouraged by an emergent central government policy allowing more freedom for developers of redundant farm buildings, to meet the demand for authentic rural housing. Thirdly, there was the continued provision of *infilling* developments *within* existing villages. Such developments were often permitted as long as they fitted within the tightly defined village envelope.

These three types of housing development are by no means peculiar to southern England. Nevertheless, it is in this region where their occurrence has been both most acute, and the planning and regulatory responses most eagerly developed, albeit *ex post* rather than *ex ante*. This is partly because of the economic and social trends in the South and the way housing provision of this kind allows for the increased incursion of middle-class residents into rural areas. Moreover, once these residents have "bought their way in" they often attempt to protect the village environment they have acquired: the so-called NIMBY effect (we go on to assess the social consequences of housing development at the villages level in the next chapter). It is our contention that the provision of rural housing has, particularly in recent years, been *socially selective*; it allows for the increased middle-class dominance of rural areas by excluding those unable to compete in the housing market. This "selectivity" is not simply an "unintended outcome" of local institutional mechanisms developed within a "managerialist" context. Rather it is the result of a process of calculation where certain assumptions are made about the workings of the housing market and others are made about the types of "consumers" this market should cater for. Needless to say, the market is deemed to be "class-blind", so those who cannot "compete" for housing find their needs neglected within the prevailing system of provision.

Belatedly, the fate of those people who lie outside this "exclusive" market is coming to be recognized as an issue of concern. During the period of our fieldwork we became aware of an increasing interest, by both planners and residents, in social housing issues in Aylesbury Vale. It was recognized that poorer families (often characterized as "locals") were being driven out of the area by rising house prices and the shortage of public sector housing. We will briefly consider below the steps taken by the local authority in Aylesbury Vale to provide housing for poorer social groups and the likely limitations to current approaches.

While the rural housing literature has emphasized the planning and policy implications of the housing land-development process, we wish to deviate significantly from this approach. We argue here that through the processes of representation, housing developments are both *socially constructed* and *mediated* by sets of strategic and local interests. Moreover, we also have to recognize that the social processes set in train during the "physical" development of housing are reinforced once successful development and occupancy has occurred, as new housing may literally accommodate new social interests and groups. Rural housing development, as Castells and Harvey have noted in the urban arena, is a highly *socialized sphere*. This sphere includes both production (through land development) and consumption (through purchase and occupancy). As our case studies will show, it is rural housing development as the social embodiment of different middle-class aspirations which, we believe, particularly typifies rural Buckinghamshire.

Housing developments lend themselves to different forms of representation at different spatial scales, associated with the construction and use of supply and demand statistics, the definition of costs and benefits of development, and the calculation of value for the producer, the consumer and the local resident. The calculations are made by strategic and corporate agencies involved in the development and regulation of housing. These are not just associated with net rates of return, but with the allocation of land supply, the aesthetic/design and architectural value of the development, and the immediate social context, i.e. the village or settlement, in which a development is to be situated.

We begin, therefore, by providing an outline of the changing regulatory context of housing development in Aylesbury Vale with reference to national, regional and local planning policy. We focus here upon how land becomes allocated for housing. Secondly, we detail and analyze three case studies of the housing land-development process, exploring the similarities and differences in each and the degree of integration these hold for the formal processes of land allocation. This allows, in conclusion, an assessment of the interaction between planning, the house development sector and local residents. The social impact of such processes at the village level provides the major focus for the following chapter.

Defining land for housing

The system of housing regulation can be characterized as a "pyramid" at the top of which are statistics for population growth and household formation. These figures provide the basic assumptions upon which calculations of housing needs in the South East are based. Further down the pyramid the provision of an "adequate" and "continuous" supply of housing is achieved through the following planning hierarchy:

- regional guidance and structure plans, which provide overall policy targets
- local plans which provide criteria for the release of sites and identify potential housing sites
- land availability studies which identify a five-year supply of sites for housing and indicate the extent to which plan requirements are being met (DoE 1991a).

However, the patterns of regulation which derive from this hierarchy must be understood in particular spatial contexts. As we have discussed

in general terms in Chapter 2, counties such as Buckinghamshire are subject to considerable population and housing pressure, despite the short-
term volatilities in the housing and construction sector. Districts in the
outer areas of the region have been particularly pressurized as county
planners have sought to channel their housing allocations into specific
areas beyond those defended by green belt and AONB designations. In this
sense, population growth has been effectively managed through the planning hierarchy, and the planning system still retains significant influence
over the siting of housing developments. This development sector is
therefore highly regulated through central, regional, county and district
government guidelines and plans. The overall result of these processes
has been to selectively target areas for new housing largely beyond the
highly preservationist and middle-class suburbs and villages in the green
belt and AONB areas of south and mid-Buckinghamshire. As we shall see,
these regulatory processes influence the operation of housing markets,
for they codify the "positional" nature of rural housing in one of the rich-
est rural residential spaces in the UK.

Strategic definitions

The maintenance of an adequate supply of land for house-building in the
South East has been a major concern for local authorities and house-
builders alike. According to a DoE consultant's report (1992a): "the supply
of housing land has been more responsive in the 1980s than in the 1970s,
because of the more rigorous requirement to maintain a five-year land
supply and improved processing of planning applications" (para. 9.33).
Current systems for assessing land availability were established in 1980
(DoE Circular 9/80) and more specific guidance notes (e.g. Planning Policy Guidance Note 3 [DoE 1984] and the subsequent revision published in
March 1992) placed pressure upon county and district councils to calculate dwelling replacements over a five-year period using as a reference
the most recently approved planning policies. For the county councils
this meant deducting the actual dwelling completions between the base-
date of the structure plan and the survey, spreading the unmet provision
over the remaining years of the plan and deriving a residual average
annual rate. A five-year implied rate could then be calculated.

This process provides the basis for the annual monitoring exercise undertaken by the county councils with the co-operation of district councils.
Planning Policy Guidance Note 3 (DoE 1984) suggested that these studies
should be regularly undertaken with the house-building industry, normally every two years. While this proposal may have been introduced
under pressure from the HBF, who were keen to get involved in the process of land release for housing, in practice, by the late 1980s the participation of the house-builders was, at best, patchy. For instance, in
Buckinghamshire the County Council co-ordinated six studies of housing

land availability. The first two were in collaboration with the HBF (1981 and 1983), while those undertaken in 1985, 1987, 1988 and 1989 were conducted by the local authorities on their own. The 1990 study was again conducted in collaboration with the house-builders in the light of a renewed pressure on local authorities to "take more account of ownership and marketability factors when assessing the availability of sites. The most straightforward way of doing this is through the existing medium of joint studies . . . listening to the advice being proffered by the house-building representatives" (DoE 1991a).

The degrees of collaboration required between county and district authorities, the house-builders and the government ensures that there is a constant dialogue between these parties, mediated through SERPLAN, the regional planning forum, as housing land requirements are calculated. The DoE issues guidance figures (having consulted with SERPLAN and the house-builders) derived from OPCS population and household projections, which are then passed to SERPLAN to apportion to the counties. SERPLAN holds the responsibility to distribute the figures to the counties, which eventually feed them into the structure plan review process. Within the structure plan the figures are divided into district targets, and it is up to the district and county officials to find sites, through the operation of land availability studies, which meet the perceived levels of "demand".

The OPCS household formation figures, which provide the basis for the calculations of "demand", are often hotly contested. In 1992 for instance, the Secretary of State for the Environment (Michael Howard) announced an additional provision of 57,000 new dwellings per year for the South East, giving an additional 855,000 dwellings over the period 1991 to 2006. He encouraged SERPLAN, the regional planning guidance authority, to look forward to 2011, given that some county structure plans were already reviewed up to that year. New housing provision for the period 1991 to 2011, at an average rate of 57,000 dwellings a year, would, it was argued, meet the projected rate of growth of households, with a margin of 3,400 dwellings per year to account for vacancies, second homes and a reduction in shared dwellings. While SERPLAN and the local authorities accepted these figures ("as a basis to plan on" – SERPLAN 1992) the CPRE saw the announcement as a "major blow for the environment and the government's credibility. [This] announcement backtracks on previous commitments to change planning policies in the South East so that the region can live within its environmental means" (press release, 9 July 1992).

The CPRE were particularly concerned that the figures would give rise to rapid development as the construction industry eventually came out of recession, given the considerable amount of land still available for development. SERPLAN viewed the figures with more magnanimity. They saw the pressure for development in the counties as largely internally generated rather than through large-scale population in-migration. The major

problem was achieving some form of regional balance, given that histor-
ically there had been a population and employment "drift" from east to
west, with all of the "Western Crescent" counties accommodating more
than their average share. However, even with this proviso it was clear
that the targets handed down by Whitehall and SERPLAN to the counties
and districts were far from excessive, given the structure plan and land
availability guidelines. The land availability studies in all of the South
East counties identified in aggregate far more land than was required
according to structure plan targets. The bulk of land supply (59%) continued to be located within existing settlements, with most of the remainder
on their peripheries; very little land supply related to rural sites. Given
the volatilities brought about by the recession, in the South East as a
whole two-thirds of the available land supply for the following five years
(1992–7) already had planning permission. This supply was more than
sufficient to meet structure plan requirements over the five-year period.
From the perspective of the strategic planners, structure plan targets and
land availability surveys were providing an adequate planning response
to housing demands for land, given assumptions concerning higher levels of infill and the growth of existing settlements over and above the
use of rural greenfield sites, and the need to strike some kind of regional
balance.

The local definitions

Two major problems were faced by the county and district planners of
Buckinghamshire. First, they had to harmonize and synchronize the inev-
itable differences in the time periods of certain structure and local plans
with the periodic pronouncements and targets from DoE and SERPLAN.
Secondly, in periods of increased economic volatility in markets and local
economies (as experienced during the study period of 1988–92) severe
gaps and discontinuities in supply and demand and actual development
outcomes can occur. In essence they have to cope with the particularities
of local space, the requirements of corporate developers, and the volatilities of demand over time. These can either strengthen or undermine the
planners' ability to negotiate with developers. In the context of Buckinghamshire as a whole (as we outlined in Ch. 2), planners have attempted
to channel these pressures to the northern parts of the county, particularly in and around Aylesbury and Buckingham. The present structure
plan, prepared in 1986 and approved in 1989, specified the target of
11,200 new dwellings for Aylesbury Vale District between 1986 and 2001,
with 8,100 being targeted for the period 1986–96. Exceeding even Milton
Keynes, this represented the largest district increase in the county.

Information on housing land availability in the private sector was produced by the district councils. Tripartite meetings between the House
Builders' Federation and the county and district councils were held in an

attempt to agree levels of house-building during the study period (most recently 1990 to 2005). Although differences were resolved in these meetings, agreement could not be achieved on some key issues, including the availability of some sites or the acceptance of large "windfall" sites as a category. The agreed land-supply study considered future housing land availability within the county arising from:

- large private-sector sites with planning permission
- large public sector sites with planning permission
- large identified sites
- small sites with planning permission and small "windfall" sites.

In Aylesbury Vale, where availability through planning permission had already been agreed, large sites predominated. These would produce 1,863 new dwellings by 1995. Large public sites – much smaller in number – would yield 400 dwellings. One of the largest private sites was the Watermead "new settlement" (see case study 1 below); also significant were Stoke Farm, Aylesbury (245 dwellings) and Cold Harbour Farm, Aylesbury (1,450 dwellings). These sites indicate a spatial concentration around Aylesbury itself, with nearby villages coming under particular pressure (see Fig. 3.1). One such area was around Stoke Mandeville road leading southwards out of Aylesbury, where 224 dwellings had been given permission along with a host of smaller sites. This area throughout the 1980s experienced considerable large-scale development pressure (see Fig. 3.2). For instance, a second new village settlement was proposed by a development company in 1988 and was refused on appeal in November 1989. The glebe land in Weston Turville, a nearby suburban village, was also to absorb a further 17 dwellings during the period, followed by a much larger proposal in 1989 (see case study 3 below).

Most of the larger sites (both public and private) were to be located in and around Aylesbury, with seven out of the nine public sites being associated with Aylesbury itself. Large identified sites without planning permission (to accommodate a further estimated 1,654 homes) were also situated in and around Aylesbury (i.e. Watermead, Stone Hospital, Stoke Mandeville). Overall, the land-supply exercise produced agreed sites for 3,141 dwellings, only 323 short of the Structure Plan requirement (see Fig. 3.1). It was concluded in the 1990 Housing Land Availability Study that while broad agreement – aside from differences of opinion about the number of windfall sites and certain large sites – was reached concerning the land-supply budget for Aylesbury Vale, the agreed levels of house-building over the period 1990–93 were likely to show a shortfall of between 475 and 3,425 dwellings. This was principally due not to the restrictiveness of the exercise but to the volatilities in the demand and supply factors associated with both housing markets and policies.

Debate continued over the calculation of "windfall" rates in these projections (currently estimated at 170 new dwellings a year in Aylesbury

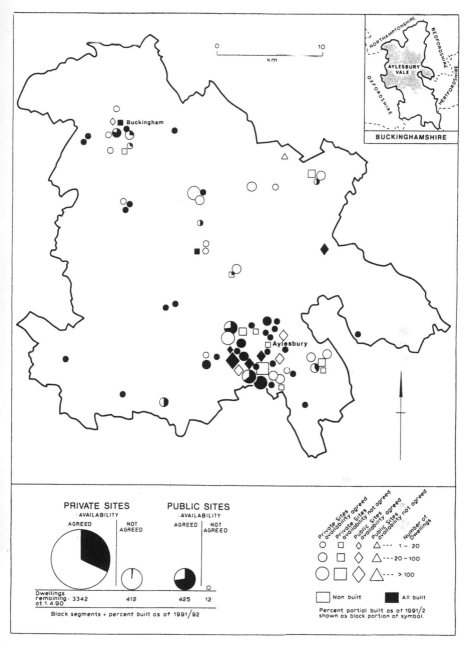

Figure 3.1 Aylesbury Vale housing land availability, 1989.

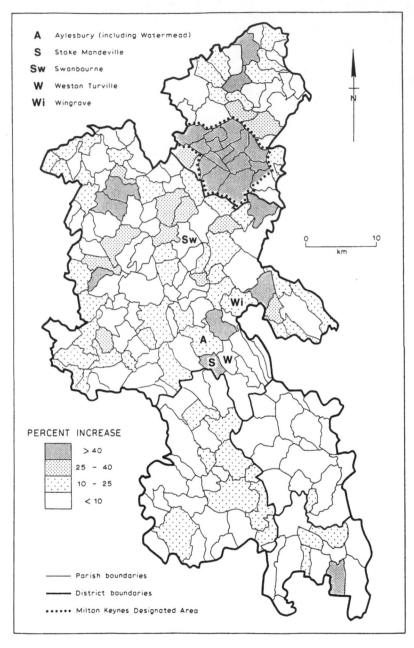

Figure 3.2 Aylesbury Vale population increase, 1980–89.

Vale). Moreover, while the large house-builders were able to codify their land-supply requirements in the Housing Land Availability Studies, many began to find these too restrictive, given the volatilities in the market and the fact that they were indeed competing with each other at the local level. They wanted a planning policy that was flexible and compliant to their particular spatial and temporal needs. Many developers felt "railroaded". In their representations to the *Rural Areas Local Plan* on 1991, landowners and developers made clear to the District Council that: "adequate allowance should be made for constraints on the availability of land imposed by considerations such as ownership and marketability". They referred to the "underperformance" of sites affected by "drop out". Aylesbury would also be reaching "its own natural growth limits" and there was likely to be "a shortfall of opportunity within Aylesbury". According to the District Council, the HBF advocated:

> increasing the dwelling provision figure in the Structure Plan by 10%, suggesting this would be a "more appropriate" method. This would build into the plan flexibility to meet "short-term revisions to the housing targets" and to allow for "drop out" in the form of non-completions on sites identified in plans or with planning permissions. Otherwise, it is suggested by respondents (particularly house-builders and landowners) – that the Structure Plan figures may not be met, especially in the latter part of the plan period. (AVDC 1992: 139)

At the local level then, the operation of land availability studies serves as a major focus for conflict and debate between the house-builders, the different elements of the planning system, the government and residents. The results of such studies remain particularly circumspect given the "boom" and "bust" experiences in the housing development sector of the late 1980s. Yet the debates occur in a context of traditional assumptions around the size, spacing and containment of settlements. In Aylesbury Vale the Structure Plan, Land Availability Studies and the new District-wide Local Plan all allocated land for building in and around the main centres of Aylesbury and Buckingham, even though the market and the development sector might become slower to exploit these opportunities than might have been suggested in the mid-1980s.

The prevailing political and economic conditions of the early 1980s may have forced land availability studies onto the planning profession, but by the end of the decade, having coped with the consequences of these, and having witnessed a housing development sector in recession, planners found an abundance of land towards which to direct developers, even though this may not have accorded with the latter's particular requirements. For developers the ability of the regulatory system to control the aggregate levels of supply was welcome, but the constraints it imposed on their flexibility at the local level were not. However, as we show in our case studies below, those house-builders who were able successfully to negotiate the planning system found rich pickings in the boom years of the 1980s.

Affordable housing

The housing policy framework has been successful in delivering private-sector housing – for instance, between 1981 and 1991 the number of owner-occupiers in Aylesbury Vale increased by 12.9% (*Regional trends* 1993). But it has been much less successful in delivering cheap, affordable housing, particularly in rural areas – in 1991 housing associations and the local authority built only nine houses in the District (*Regional trends* 1993). However, the demand for this type of housing has become increasingly apparent at both the national and local levels. The government issued guidance on the rôle that the planning system could play in securing provision of affordable housing (DOE 1992c). This pointed to a more significant rôle for land-use planning in developing instruments to encourage low cost homes. SERPLAN now completes annual surveys of "affordable housing" for the South East. Such provision is defined as: "New housing provided with a subsidy to enable the asking price/rent of the property to be lower than the prevailing market price/rents in the locality, and which is subject to arrangements that will ensure its availability in perpetuity" (SERPLAN 1992: 2). The subsidy may be financed from public funds, such as central government grant, through local authorities under the housing legislation, or from a modification of land values. It may also be funded by the private sector through, for example, the internal financing of schemes.

SERPLAN's studies began to show an increasing proportion of "affordable" housing construction. Within the ROSE the proportion was 20% of completed dwellings in 1990/91 compared with 12.5% the previous year. There were, however, marked variations between the counties. Supply was allegedly greatest in Buckinghamshire, where more than a third of all completions were defined as "affordable" dwellings. However, a large proportion of this was the result of development activity in Milton Keynes and other urban areas. In no southeastern county was the supply of affordable housing in 1990 sufficient to meet the needs of the younger, newly forming households. The sale of local-authority and new-town dwellings over the preceding decade represented a significant loss of affordable housing stock. Throughout the decade, in most counties the number of dwellings which had been sold far exceeded the new provision of low-cost housing. The sale of local authority housing had tailed off by the early 1990s, such that the construction of low-cost housing was beginning to exceed council house sales by 1991. Nevertheless, "right-to-buy" policies severely exacerbated the problems of affordability. Over the nine-year period 1980–81 to 1988–89 there was a net loss of over 70,000 public-sector dwellings in the ROSE and a regional loss of 220,000 dwellings from the public rented sector. As a consequence many of the region's planning authorities (including Buckinghamshire) began to formulate policies designed to secure affordable housing. However, local planners were still reluctant to define "need" and allocate potential land specifically for this purpose. In Buckinghamshire, although some of the districts are now including these policies in their emerging local plans, the current Structure Plan does not include an "exception" or a quota policy for affordable houses

In Aylesbury Vale, despite the growing concern of planners to calculate levels of "need", there remains at the time of writing a major crisis of access. There are no ready assessments of housing need (only five out of the 101 parishes have conducted surveys on this topic). The council waiting list is restricted to those who have been residents for at least 12 months and who earn less than £11,500 per annum. High rural housing prices, in addition to restrictions in the provision of council housing, have intensified the problems of access for people on low incomes. By 1990, in many of the villages house prices had risen above £60,000 for *all* properties. Aylesbury Vale Council has, however, made some attempt to supply low-cost housing in the context of limited resources. By 1991, 21 parishes out of the total of 101 were selected as appropriate for small social-housing schemes. In the majority of cases these developments represented partnership arrangements with the private sector using the local authority's land bank. This method of provision has produced nearly a thousand new homes which have been sold at discounted prices to Council nominees. One consequence, however, is that the Council's land bank is now largely depleted. As a result the local authority is exploring alternative methods of providing this type of housing. These include:

Developing a council-shared ownership scheme   The district council, at the time of writing, had just completed a rural shared-ownership scheme of 18 dwellings, where homes were sold on a 50% equity stake ranging from £31,500 to £37,000. Other small schemes are planned, but these will only be limited. This is occurring within the context of no council house-building anywhere in the district after 1991/2. The council alternatively began to establish partnerships with private builders. For instance, in 1989 six properties were bought in the village of Haddenham which were then resold on a 50% shared ownership basis to Council nominees. The Council also acquired a stake in the equity of homes where existing owner-occupiers would otherwise have become homeless (due to marital breakdown or mortgage default). This left the original occupiers with only a 29% stake.

Offers of land to housing associations   There was relatively little activity by housing associations. With growing encouragement from the Council, some associations began purchasing previously privately owned homes from developers, to be let on assured tenancies to homeless families. The Council provided £500,000 from its housing programme, giving deliberate priority to rural areas. The National Agricultural Centre Trust (NACRT) was involved in survey and development work in five villages, but no firm proposals materialized.

Landowners and developers   Overall, as we shall see below, landowners and developers show only a passing interest in the provision of "affordable" homes. Where they do, it could be argued that their interest in affordability is only to the extent that it is a "sweetener" for larger private-sector development proposals. However, in some cases Aylesbury Vale District has been able to make some land available explicitly for developers to establish low-cost homes. In the village of Brill, for instance, public land was made available on the north side for a developer to build 36 joint-venture homes (mainly one- or two-bedroom units). These were mostly bought by Council nominees. The development lies just on the edge of the village conservation area, and the high density has attracted some criticism from local residents. This has not avoided the problems of housing access in the village. When such joint ventures have been developed (there are currently 16 villages which have small-scale joint ventures providing 310 homes), demand for the properties has been intense. For instance, for the 16 new homes built under these schemes in the village of Stoke Hammond (prices from £35,000 in 1987) there were 120 nominees.

Despite high levels of need, the provision of affordable housing has been limited in Aylesbury Vale and looks set to remain so in the near future. Private-sector owner-occupied housing has dominated the housing sector, particularly in rural areas where some of the most exclusive housing is located. In the rest of this chapter we turn to examine development case studies which illustrate the types of housing being delivered by the market and the planning framework. First we examine a development that bears some resemblance to a new settlement, particularly in terms of design, layout and house type. Although this development was not situated in open countryside, it fulfils many of the functions proposed for new settlements in the 1980s. Secondly, we examine developments within villages, looking first at up-market barn conversions in a desirable village location and, secondly, at a large "infill" development in a suburban village. The latter case gives us an example of how affordable housing can be subsumed into larger private-sector proposals.

Housing case studies

Case study 1: the new settlement

The issue of new settlements pushed to the fore the whole question of social change and its impact on rural areas. In 1983 Consortium Developments released proposals to build up to 15 new self-contained, "balanced" communities in open countryside within 30 miles of London, thus relieving the pressure on other rural towns and villages in the South East, in a private sector equivalent to the new towns programme of the post war years (Ambrose 1986). According to their advocates, such as Nicholas Ridley, new settlements could not only soak up a considerable amount of market pressure but "may also offer the opportunity for better standards of housing design and better environmental quality than the more conventional types of estate development" (*Planning*, 14 July 1989). No new settlements of this type have been granted planning permission in Buckinghamshire, although proposals have been submitted to Aylesbury Vale District Council. According to the Council: "the strategy of the Structure Plan provides for development at and within some settlements . . . . It does not propose the establishment of new settlements. The District Council is satisfied that the dwelling figures in the Structure Plan will readily be achieved in accordance with that strategy and there is no need for new settlements in Aylesbury Vale nor at this stage identification of other locations for development outside existing built up areas" (AVDC 1991a: para. 3.20). In responding to specific development proposals the Council has consistently argued that it would grant permission for development only if it was required to meet the essential land needs of agriculture, forestry or related industry, particularly as existing settlements could accommodate additional growth.

However, in the mid-1980s the position was more problematic, for the Council found itself unable to meet the requirements to ensure that sufficient land was available to satisfy housing needs in the structure plan. To remedy this the Council gave permission for 800 dwellings to be built on a self-contained site on the northern edge of Aylesbury. Essentially this was a new settlement, but instead of a countryside location it was situated in the urban fringe just outside the existing town.

The origins of this development can be traced back to the beginning of the decade when a house-building company named ROYCO were looking around Aylesbury for suitable development sites. ROYCO was a "niche" house-builder based in the South East, specializing in "quality" housing estates. It was not an especially large firm: it had a turnover of around £20 million per annum and on average built 150 houses a year.

In 1980 AVDC produced a planning brief the main purpose of which was to determine whether a ring road should go to the north or the south of Aylesbury. ROYCO concluded that a northern route was most likely.

75

The company then took out an option on the land which would eventually become Watermead. A ROYCO Director lived locally and knew the landowners socially through the Round Table. The land itself was close to the northern boundary of Aylesbury, and the landowners were clearly looking for some way to realize its development potential. They considered that ROYCO might be able to achieve this and so allowed them to conduct all the negotiations on their behalf with the planning authorities. The option covered 81 hectares (ha), of which 48.5 ha were low-lying land subject to flooding. At this time ROYCO approached the planning authority with a proposal to construct lakes in the low-lying land, with parkland to be made available to the local authority. In return ROYCO sought planning permission to build on the remaining 32.5 ha. At that time the building land was calculated to be worth around £500,000 per ha, yielding approximately £16 million to the vendor, who would then stand the cost of building the amenities. However, the planners rejected the scheme.

The scheme was resurrected in 1984. At this time there was a change in personnel in the planning department and ROYCO approached the new planning officer with the much the same proposal, although there was one modification: the spoil from digging the lake could not be transported off site, so a ski slope would be constructed. This was seen as an additional amenity resulting from the scheme. On this second occasion the proposal received a much more positive response. The reason for this change in attitude lay not just in new personnel but also in the fact that AVDC was in a difficult land-supply position and was unable to meet its five-year housing land-supply targets. The planners now saw this development as a significant way of easing the county and district housing targets, while minimizing opposition from the local population by concentrating the housing in a self-contained settlement.

Negotiations now began in earnest. The planners had several provisions which they wanted included in the new proposal. First, they made clear that they would agree to the size of the housing development (32.5 ha with just under 25 houses per ha) only if a legal agreement was included which phased the development to 100 houses per year. They felt that granting permission for all 800 houses (the full amount permitted) at once would allow ROYCO to dominate the local housing market. Secondly, they wanted the leisure facilities to come "on stream" at the same time as the first batch of houses rather than at the end of the development. Thirdly, they required a covenant on 32.5 ha behind the development site to act as a "green belt" preventing any further development close to the village of Bierton, which is close to both Aylesbury and Watermead. Lastly, the planners wanted some large "executive" type houses in the project in order to relieve some of the pressure on the villages in the area. They clearly saw this project as having the capacity to solve a range of problems. Outline permission was granted in September 1986.

Having gained outline permission ROYCO started to prepare a more detailed scheme. They had recently begun to specialize in "concept" housing developments. The first "value added" scheme, undertaken by the company in 1980, was 17 houses in a gated complex with cobblestone roads. According to a Director of the company: "What we tried to do was create an image of a traditional London square. Basically the concept was that if you go to Spain or somewhere they do these villages that are all the same stylized form and that wasn't happening here so we thought that we'd do one. This first one was very successful and it led us into trying to do something that was different with our housing". The company then moved to a larger scheme in Sittingbourne, covering 40ha of housing. Once again this was a highly "stylized" development, as the Director explained: "We did it in sort of colonial style. We started out with a village green and a duckpond, putting English country cottages round it"

Aylesbury Vale councillors and Council officers visited this latter scheme during the negotiations over planning permission and insisted that the company raise the standards of housing design for Watermead. Therefore, the company put much emphasis on the design stage. The design features of these developments became a major issue in their marketing. Although the firm recruited outside architects, the design was very strongly led "in-house", specifically by the Director. He explained the process in interview:

> We do a lot of photographing of nice features, and our style is we try to create images of what people think. I've got this theory that if you hear a record once you think it's pretty awful, if you hear it a hundred times you think its great. There's a lot of style in architecture that people hardly notice but it is constantly there and there's a lot of it in London . . . what we do is go round and photograph features . . . and say this is what we'd like to develop. Our houses look more like a London square than a London square looks.

Having "collected" such design features, these are then fused into a development model. The heterogeneous elements begin to evolve into "their own style" so that "we're not trying to follow any general trend but we are trying to build something which is ageless". Having constructed the model "then we video the model because we feel that by looking on a TV screen it comes up in a form you can understand". Film and television images seem to inform the very design itself, not just the way the designs are presented: "Our schemes are deliberately like film sets. We think that the trend is that people watch a lot of television where colour television creates a tone because the lighting is different, you never see a true colour, you see a technical or bright colour. What we try to do in our house styles is create that same image, that same glamour"

Watermead was designed around the lakes and this provided the opportunity to present the project as a lifestyle development incorporating health and glamour. The Director said: "We reached the conclusion that Aylesbury was probably looking for an obvious style, a healthy lifestyle – water activities, jogging track, cricket square etc. This was basically the concept – two lakes to maximize the lake edge for housing purposes and we wanted space for a wildlife and conservation area". This heavy design emphasis was not confined to the external aspect and the overall environment; it also included the interior of the houses. Again, every effort was made to recreate images that are familiar: "What we do is get a glossy magazine, we look at the ads in there and try and create them . . . Once again its this feeling that you see a picture of say a kitchen, if you can go and see that identically replaced because you've seen it before it clicks in the mind" Every aspect of this development was to be fused into a conceptual whole. Although images and styles were to be borrowed from a variety of sources the overall aim was to create a "village" with a "timeless" quality. It is also clear that the initial designs were made with some consideration of how the scheme would be marketed and who the marketing would be aimed at. As the Director put it:

> We do look at our design in terms of how they will photograph because its vitally important that you get your photographs marketed. When building a new house its got no life in it, so the styles we're looking at do create a picture which we need to sell. We've constantly found that people are surprised that when they get to the site it looks like the photographs. And it's because the sites look like a film and the film looks like a site – it's all a bit glitzy

The detailed negotiations were conducted with the planning authority on behalf of the landowners by the company Director. He secured the landowners' agreement to all the planning provisions. Most importantly, he secured their agreement on the provision of the amenities. The leisure and parkland facilities were to be financed initially by ROYCO, but would be recouped from the first payments made to the landowners. Given that the cost of these facilities would be £4–5 million, that the land release would be phased and that they had an option agreement with ROYCO which tied the price of the land to current market prices, the landowners would in effect receive nothing in the first 1–2 years. It is perhaps not surprising that the company's Director felt all parties came out of this well:

> Because of the deal we were able to put together with the landowner, the landowner came out of it laughing because we'd created 80 acres [32.5 ha] of housing land which today is probably worth £40 million, so the landowners' got his £40 million – subject to tax – and it cost him £5 million to get it and he's very well pleased. Had he been granted permission anyway there's no way, out of £40 mil-

lion, that he would have spent £5 million. He may have been phil-anthropic but it wouldn't have enabled this sort of thing to happen. . . . And everybody's come out a winner in my view. We've done well because we are developing a nice environment, the [Planning] Committee has done well because they've got parklands and things they wouldn't have got.

This was borne out by the other parties. As one of the landowners said: "As a commercial consideration we paid more than we needed, but both of us wanted the development to be an asset to the area and were keen to see recreational facilities as a real asset". According to the District Planning Officer: "it is highly unlikely that this project would have come to fruition without this approach by the landowners" (quoted in Elson 1989). And the local council felt it too had achieved a reasonable deal, as a councillor explained: "Watermead is seen as a prestigious development for Aylesbury Vale (by both officers and members). Watermead gave us the opportunity to provide executive housing in Aylesbury – took the pressure off the villages and the staggered land availability was evolution rather than revolution"

Construction began in March 1987, starting with the lake and the ski slope (because of a wet summer in that year the ski slope slipped and did not attain its full height, but was still eventually operational). The lake was designed specifically for watersports and in particular to accommo-date jet skiing. Another lake was added to allow, according to ROYCO publicity material, "more leisurely Edwardian pursuits, such as punt-ing". The first batch of houses were built around the lakes. These were the large executive houses, at the top end of the price range.

Having laid the groundwork and put the main elements in place, ROY-CO then started to market the scheme. Watermead was designed in an extremely image-conscious fashion and the marketing of the scheme cen-tred around this aspect. The first publicity brochure declared the devel-opment to be "the perfect location for a way of life you previously only thought available when you are on holiday . . . . When you approach Watermead via the long tree-lined boulevard, you feel as though you are entering another world . . . . At Watermead your home is also your holi-day retreat". The housing style was now characterized as "Edwardian" and the development "re-creates all the warmth and charm of a tradi-tional Edwardian village . . . Watermead uniquely combines sporting and leisure facilities with a delightful modern village comprising beauti-fully designed Edwardian country cottages and detached villas . . . set amidst acres of beautiful parkland and lakes". While the overall feel may be "Edwardian", the houses themselves are not subject to the whims of the fashion conscious where Edwardian houses may be "in" today and "out" tomorrow: "The 'timeless' nature of their design means these properties are unlikely to be affected by trends or fads and as a result they offer tremendous long-term investment potential, as well as providing a better quality of life"

The "quality of life" theme was prevalent; the brochure contained photographs of windsurfers, skiers (apparently on some Alpine slope as opposed to a dry ski slope in Buckinghamshire), joggers and various exotic birds on the lake. Furthermore: "the lakeside theme and the stunning beauty of the environment are also reflected in the street names. Imagine telling your friends you live in Kittiwake, Sandpiper or Osprey Walk". The "completeness" of the "village" was rounded off by the range of on-site facilities, the "distinctive village square, with its pink and cream-painted pub, restaurant and shopping mall, set around an attractive piazza . . . you have all the amenities you could ever want".

The marketing campaign was initially very successful ("clever hype" according to a District Councillor we interviewed). The first houses to be built were sold before completion. In 1989 the development won the *House-builder* "What House" award. Everything seemed to be running fairly smoothly. However, the onset of recession soon slowed the rate of progress and by the beginning of 1991 only 170 houses had been completed, far fewer than the available permissions. According to the agreement reached with the landowner, ROYCO was committed to taking 4 ha a year at the prevailing market price. However, with the recession beginning to bite, the company found this obligation hard to meet (at one point it had £1 million worth of unsold houses on site and prices generally fell by about a third). Plots were therefore sold to other house-builders such as Bryants and Admiral Homes. The next stage of the development, Watermead Gardens, designed around the second lake, was postponed because of the depressed market, but eventually got under way in 1992, with the emphasis on first-time or single buyers and one- and two-bedroom apartments.

Furthermore, there were problems with the leisure facilities. The planning gain package stipulated that the leisure facilities should be available to the public. However, Watermead was designed and marketed almost as a private village. For instance, although the development was ostensibly an offshoot from Aylesbury, there is only one entrance to the site ("via the long tree-lined boulevard"); reaching this entails driving out of town, clearly distinguishing Watermead from Aylesbury itself. The sports facilities brought people onto the site and were extensively used. Particularly popular was jet skiing on the lake. However, this is a noisy activity and it rather upset some of those who purchased the expensive waterfront houses. Complaints were made and ROYCO, as manager of the estate, was forced to try to curtail the amount of skiing. This resulted in an acrimonious row between the company and the subcontractor providing the facilities, and led to some unfavourable publicity in the local press. Eventually an agreement was reached whereby jet skiing was restricted to two sessions per week. However, complaints continued to be voiced in the local press and the lake sports contractor responded by saying that "it is obvious that one or two residents . . . will not be content until the public access to Watermead facilities is denied" (*Bucks Herald* 17 October 1991).

The overall concept was clearly aimed at a particular social stratum. As the ROYCO Director explained: "It's social engineering in a sense – I don't mean that deliberately – all we are doing is designing the environment. But we do have a lot of pretty girls, it's that sort of place, there are the smart young couples driving white GTi's, its that sort of place". When the facilities first opened it was, according to this Director, "definitely the place to be, particularly if you were in the white GTi set". However, the pub acquired a new landlord "who has moved it down-market". The clientele no longer "fits" and further bad publicity has resulted, with newspaper headlines such as "Men Fought in Lake" (*Bucks Herald*, 15 November 1990). The Director concluded "it is difficult to get that balance right to create the sort of village that we'd like it to be" Nevertheless, four years after it was established, a village atmosphere was beginning to evolve. According to one resident: "although the houses are all new it is like a village in the sense that everyone knows everyone else and what they are doing"

Watermead was included in our social survey, with 58 households interviewed. Of these 26% had moved to Watermead from Aylesbury, 31% from elsewhere in Bucks, and 41% from outside the county. Seventy per cent of the Watermead respondents were in the 25–44 age group. The socio-economic groups of the respondents were as follows: group 1 8.6%; group 2 39.7%; group 3 24.1%; and group 4 5.2%. There was a significantly higher ratio of respondents in the top three groups than in the sample as a whole, and fewer in the economically inactive group (22.4%). When asked whether they thought they lived in a town or a village, 56.9% said a town and 43.1% said a village. Their main reasons for moving to Watermead were as follows: 29.3% said housing type; 22.4% mentioned work; 20.7% said the "environment"; and 20.7% cited marriage. Finally, in response to the question "what is the best thing about living in this area" 50% cited scenery/environment/countryside. In short, Watermead tended to attract young people in the top three socio-economic groups, partly through the distinctive housing and environment it offers. Having reached Watermead, however, the level of commitment to the place was not particularly high and there seemed to be some confusion as to whether Watermead was really a village or part of a town.

In what sense can we characterize Watermead as "rural"? Clearly an attempt was made to turn what would otherwise have been a fairly self-contained housing estate into a "village". This entailed the construction of a particular kind of environment incorporating "lifestyle" features, such as leisure facilities, and environmental features, such as the lakes, conservation areas and trees. All this was welded into the Watermead "concept", an "Edwardian village", where the constructed environment and the houses merged ("the properties are in perfect harmony with their surroundings" – ROYCO publicity material), giving Watermead a distinct identity.

This merging of a variety of "stolen" elements leads to a form of "manufactured" rurality which has only been partially successful, given that many residents do not seem to regard the development as a rural space. However, an "idyllic" notion of village life certainly gives these elements a sort of spurious unity, allowing Watermead to span the urban and the rural. Symbolically, the village "looks" out into the countryside and away from the town; it tries to ignore the existence of Aylesbury as far as possible. The "brutal" modernist developments of the 1960s, which tend to dominate Aylesbury, stand in stark contrast to the playful pastiche of traditionalism represented by Watermead. But more than this, it tries to turn its back on certain elements of Aylesbury's population, i.e. those people prone to fighting in lakes.

The Watermead estate seems to indicate that "sympathetic" development, incorporating "quality of life" characteristics, is not just about building houses. It is about the "total environment" in which the houses are situated. This entails some kind of seamless meshing between the private spaces within the houses and the public space surrounding them. The public space is, of course, the most difficult of the two to manage, as the conflicts over the pub and the lake demonstrate. What is required here is for developers to do more than just build the houses, for they still play a rôle after the houses are built in arbitrating between disputes on the site. This could be characterized as a privatized version of the local authority housing estate, although developers will attempt to keep their involvement to a minimum. Their obligations are circumscribed by a series of legal agreements with all other parties, such as house purchasers, the local authority, the landowners and the contractors providing the watersports. Whether the developers can satisfactorily play this rôle remains to be seen. If they cannot and, for instance, go out of business, the question must be posed: who takes responsibility for estates such as Watermead?

The development provides us with a graphic example of how developers are utilizing certain conceptions of rurality to meet the aspirations of those seeking a non-urban life-style. That this is a kind of "mock" rurality does not seem to concern the residents of the development; they simply want to live in more sympathetic surroundings than those which prevail on other housing estates in the area. Without the provision of Watermead their only access to such surroundings would be within the older, established villages. However, the restraints on development in such places makes access difficult. Watermead clearly functions to relieve pressure on the villages and in that sense can be seen as a surrogate for a much more "authentic" rural lifestyle.

Case study 2: barn conversions and the re-creation of the rural
Watermead provides us with one significant example of how a developer attempted to "capture" the growing consumer preference for a "total" environment which provides a certain "quality of life". In the past, access to this kind of "quality of life" has often been sought within traditional rural villages. While the latter do not usually encompass the range of integrated features found at Watermead, they are, as we shall see, being forged in the likeness of an image which, on the whole, reflects the aspirations of those with the material and cultural resources necessary to reside in such places. The conversion of farm buildings to housing is one element of this refashioning of rurality which aims to recreate "authenticity" in Buckinghamshire villages. Such an aspiration can be considered to be in sharp contrast to that which prevails in Watermead where the fusion of "stolen" elements seems to be almost a celebration of the "unauthentic".

Aylesbury Vale District Council published a policy document *Conversion of traditional farm buildings in open countryside* in 1989. No explicit justification for the policy was provided, although the introduction to the Report states: "The Council, in common with many others, has been receiving a significant number of planning applications for the conversion of farm buildings for alternative uses [and]. . . . Central Government advice urges that new uses should be found for redundant buildings in the countryside. This policy document seeks to develop that advice at the local level" (paras 1.1, 1.4). The policy was not confined to any particular conversions, but to strict building types where it was seen as important that the building be of "visual architectural or historic interest" and that it "contributes to the character of the open countryside" (Policy F.1). The suitability conditions for conversion were limited to a redundancy qualification: "The building should be redundant. In a strictly financial sense a building becomes redundant to a farmer when the costs of repairs exceed the value of the building to the farm . . . . In assessing whether a building is redundant, consideration will be given to whether it is superfluous to the needs of the farm or can no longer be used normally, practically and profitably for the agricultural operations of the farm on which it is located" (para. 4.6). The Council argued that it would welcome the continued use of the building where possible but was happy to see proposals for residential or employment-generating uses since these were deemed acceptable by central government (see also Ch. 9 on industrial conversions).

This discussion document was ultimately incorporated into the *Rural Areas Local Plan*. Here, a greater commitment and enthusiasm for conversion is evident. It is argued that barn conversions "can help to reduce demands for new buildings in the countryside, encourage new enterprises, and provide jobs and housing needed in rural areas" (AVDC 1991a: para. 4.40). Furthermore "in order to both help conserve the District's countryside heritage and sustain the rural economy, the Council wishes to safeguard their character and exploit their potential for continued use by allowing the conversion of such buildings" (para. 4.41). Thus, the policy seems to embrace the concerns of a variety of social actors – developers, farmers, landowners, conservationists and rural residents (both potential and actual) – by using sympathetic development to "conserve" valuable elements of the rural landscape. In summary, therefore, the policy encourages the re-use of traditional buildings in order to "safeguard their character and exploit their potential" (para. 4.41). Modern farm buildings, on the other hand, "intrude upon" and "detract from" the landscape and "owe their existence to an essential need which no longer exists" (para. 4.42). However, confining conversions to traditional buildings, and emphasizing the use of traditional materials (blending in with the building's structure and immediate environment) is likely to raise both the cost of the conversion price and the price to the consumer. Thus, traditionalism and authenticity reinforce "positionality" in the housing market. In the rest of this section we examine two examples of the residential conversion process which are both in the same village (the village itself – Wingrave – is studied in more detail in Ch. 4).

Unlike new estates, which are generally standard high-density housing, the conversion of farmsteads allows development in a manner which pays due heed both to the original buildings and the surrounding environment. In order to understand a little more fully how this type of rural housing is produced, we can examine in some detail one such development on a site in the village of Wingrave. This site was developed in two stages. The first stage, called the "Essex Yard" development, was initiated in the mid-1980s when the owners applied for and received planning permission for four houses. At this time the buildings on the site were old store sheds and a coach house, all disused. Having been built by the Rothschilds – the former estate owner – these were listed buildings, so any conversion could not alter the exterior to any great extent. The site was purchased from the owners by a local builder who specialized in up-market, high-quality conversions. The firm was essentially run by one man who since 1970 had been reconstructing timber frame buildings in rural areas. All his business was local and he had never been interested in expanding the scale of operation. He employed only one man full-time and he undertook all the work apart from plumbing and electricity, which were subcontracted out. The builder prided himself on doing "sympathetic" conversions which retained the original character of the building:

> I hate to see these so-called "barn conversions" where they come along and they just fill it up with concrete blocks, plaster it in the old traditional way and all that sort of stuff. There's no need for it at all because what we do is just leave the main structural columns and tail beams in and put new timber stud work in and insulate it and reinforce it with plywood and all the rest of it . . . try and make it look as original as possible

Having acquired the capital to purchase the site (from "a friend who runs a lot of companies") the builder altered the scheme slightly and renegotiated the permission to build five houses, four with two bedrooms and one with three. Having obtained permission, the builder then tried to involve the local community in the scheme by inviting members of the parish council to the site to explain what he intended to do: "We tried to get on a good footing with them straight away"

The conversions were completed in 1988, by which time the first three (two-bedroom) houses had been sold for around £70,000. The last two were sold in 1988 for £100,000 and £120,000 (the person who bought the latter had resold within the year for £180,000). All were bought by people from outside the village. These were timber-framed buildings using mostly original materials, with only dormer windows and roof lights added, constructed around a courtyard. It was a secluded, enclosed development.

As this scheme was nearing completion, the landowners were attempting to obtain planning permission on the other half of the site. They put in a proposal to convert the two existing buildings on the site and to construct nine new dwellings in a cul-de-sac. This proposal was vehemently opposed by the parish council, the district councillor and most of the people living adjacent to the site. The main concern was the loss of open space and the possibility that the houses would simply look like developments everywhere and would not, therefore, be "in keeping with the village". The scheme was refused by the District Planning Committee.

The landowners then contacted the builder of the first conversion and asked him to propose "a more acceptable development plan" and then to negotiate on their behalf with the planners. As he had already established links with the local community and been responsible for what was seen as an acceptable development, the landowners felt that he had more chance of satisfying local amenity interests and the planners (particularly as his first scheme had just won an award from a local conservation group, the Friends of Aylesbury Vale). Initially there was no guarantee that he would get to build the scheme were it accepted; he was simply to be paid a consultancy fee for the planning negotiations. A new proposal was submitted in 1989 for five houses (one converted from existing buildings) in a courtyard complex, so there would be no road and less open space lost. The design of the scheme was governed by one conversion; the new houses would be constructed in the same "original" style.

There was still a considerable amount of local opposition to this proposal. The District Councillor and the Parish Council objected on the grounds that there would be a loss of open space and the development would "destroy the rural nature of the land". Nine letters of objection were received by the planning department. However, these were offset by eleven letters of support which all pointed to the "high standard" of the development on the other part of the site. Planning permission was recommended by the officers and granted by the Planning Committee.

Having acquired the permission on the strength of the builder's reputation, it would have been somewhat unjust if he had been denied the opportunity to develop this conversion. The builder believed that "if the price of land had still been rocketing they [the landowners] might have sold to the highest bidder and I wouldn't have been able to get involved". However, the price of land was not "rocketing", as the downturn in the property market was well under way and the builder entered into an arrangement with the owners whereby he built the houses and each party took 50% of the sale price after all fees were paid. The development gain was equally divided between landowner and developer. Once again the development took place in the "traditional style" (so that the houses *looked like* barn conversions) and the properties were priced at the upper end of the market. The first one to be built was sold for £165,000, having cost £80,000 to build. The landowners and the builder–developer therefore stood to gain over £40,000 each from each conversion. It was highly unlikely that any of these houses would be purchased by anyone living in the village.

Another redundant farm in the village was also converted to housing in the late 1980s. In 1987 permission was granted for the subdivision of a farmhouse into two dwellings and the conversion of three barns into three houses. In a further application in 1988 the barns were now to be converted into five houses, with some garages attached, around a floodlit courtyard. The design of the buildings and the use of construction materials was to complement the existing features and the earlier conversions. The site would include seven houses in total. There were objections from the Parish Council on the grounds that the increased number of dwellings would result in a decline in "environment standards" and have a significant impact on traffic. The District Councillor for the village considered that eight new dwellings (the proposal had been amended to seven) would amount to overdevelopment and "would result in a down-market mews development containing all the impediments of modern life" (AVDC Planning File). The Planning Office was sympathetic to the development which, it argued, had "been designed to ensure the beneficial use of these buildings which contribute to the character of this part of the Conservation Area". Permission was recommended and granted.

This type of scheme is characteristic of the way the old central part of Wingrave village was redeveloped in the 1980s. These developments represented a kind of "rustification" of the village landscape with middle-class commuters dwelling in "traditional" rural buildings. This is ostensibly a type of development in which the preservation of form conceals a transformation of function (Kneale et al. 1991). The conversion of these farmsteads to housing had a pronounced effect on the social fabric of the village. The prices that these buildings were commanding in the mid- to late 1980s put them out of reach of all but upper income groups prepared to pay for an exclusive slice of the "rural idyll". Furthermore, all these conversions were "clustered" in the centre of the village where all the traditional features could be found. The sympathetic conversions blended in with more traditional elements. However, the dominance of the middle class in the "heart" of the village was increased.

Case study 3: developing the diocese

New settlements or the more gradual conversion of former agricultural buildings represent a reconstitution of the metropolitan countryside. Our third case relates to a rather more predictable land development proposal at the heart of a village just south of Aylesbury, named Weston Turville. It exemplifies the issue of village infilling and the growing problem of affordability.

One of the traditional features of many Buckinghamshire villages is a large village green surrounded by what were originally fragmented settlements. In a few villages the settlement pattern is still fragmented around an extensive common land area of up to 10 ha. Weston Turville is such a village, located within the hinterland of Aylesbury. In May 1989 the Oxford Diocesan Board, acting on behalf of the church, submitted an application to develop a housing estate on half the area of the village glebe land. The reasons given for the development application were: (a) it was necessary for the continued benefit and development of the community and (b) the development gains were needed in order to pay for the escalating costs of the clergy. This point has been a critical concern for the fund managers of the church and it indicates the potential pressure on rural land owned by what has traditionally been regarded as a steadfast supporter of rural continuity and tradition (see Hamnett 1987b, ACORA 1990). In the application the Board's agents argued that the site "is almost unique amongst the villages of Aylesbury Vale, in that it comprises land which is wholly within the built-up area of the village, forming part of an area which has already been entirely surrounded by development . . . . In its existing state, the site has very little amenity value and cannot be seen at all from any of the main through roads in the village. The land is in fact of very little beneficial use to anyone, as it stands at present" (AVDC Planning File).

In preparing the application and, indeed, in making it more viable, three other agencies were involved in the project. First, the NACRT, with a remit to establish low-cost housing, was to allow local people to participate in developing ten houses for sale at 40% of their eventual full market value. These houses were to be specifically designed for local residents with low incomes in need of accommodation. Secondly, the Parochial Church Committee were to be the principal managers of the development, answerable to the Diocesan Board. Thirdly, Huntingate Homes Ltd were to be the private construction company responsible for building houses covering 1.5ha (half the site) at a density of 30 houses per ha.

A much smaller housing development taking up a minor part of the glebe land had been permitted a few years earlier, but the second proposal was far more ambitious in scope. Apart from the houses, it included a new village green in the centre of the estate and a new parish hall/village centre. Given that the earlier but much smaller application had gained acceptance on appeal, the Diocesan Consortium seemed confident of success. However, by the end of 1990 – one year later – the composite application was withdrawn due to "the strong weight of public opinion against the proposal" (AVDC Planning File).

Evidence from interviews and documentary sources suggests that the development project was deliberately made *composite*; indeed, it was an attempt to reproduce rural conditions, to provide social needs and to protect a much more limited amount of green land. The composite nature of the proposal was an attempt by the landowners and developers to demonstrate their social responsibility. This was deemed necessary because the land was located at the centre of an already suburbanized and expanding village, with a large population and with a local climate of hostility to new rounds of housing development. The development proposal directly attempted to recruit supporters through its provision of facilities to meet local needs and amenities. The objective of proposing the development of 10 low-cost homes represented an attempt at least to recognize the existence of the problems of access to low-cost homes for local people, particularly young families. Secondly, in a technical sense, the development proposal could have been considered as village infill. As we discussed earlier in relation to the land availability studies, in terms of land-supply monitoring and planning the village of Weston Turville could potentially absorb more infilling. The developer's application stressed the contemporary growth of the village, pointing to the fact that a considerable amount of new development had been allowed and that, despite its size, the proposal could be characterized as "infilling". By including a new village green and a village hall, the developers were attempting to reconstitute part of the village, creating both residential and recreational space for the benefit of existing residents.

However, the establishment of over 60 new homes would undoubtedly have further changed the social character of Weston Turville and, for existing villagers, have reduced the common rights of access to the glebe land, despite the developer's intention to provide a new "village green". An action group was established and it began to orchestrate opposition to the proposal. Although this development contained many features apparently beneficial to the village, it was vehemently opposed by many residents, particularly those who lived on the piece of glebe land which had already been developed. It was these residents who launched the "Save Weston Turville" Action Group. Once they had achieved their "positional" housing status they needed to protect this by fending off further development. The Parish Council meeting which considered the application had over 200 in attendance. The Action Group launched a petition and claimed that 98% of village residents were against the development. The Planning Office received many letters of opposition. Selected quotes from these exemplify the arguments used in opposing the development and also throw some light on residents' attitude to the past development of the village and the value of the remaining "green space":

Contrary to the view put forward by the applicants, i.e. the land is "not much used", I, along with many villagers cherish it as a much valued open space – psychologically and spiritually valued. It offers a variety of recreational choices: dog walking, horse riding, children's fun space, fresh air, social meeting place and more . . .

. . . the land . . . is not situated on the edge of a built up area, but is in the centre of the village, completely surrounded by a large proportion of the houses in Weston Turville. As such, it is a major attractive feature of the area and indeed is the centre and heart of the village, an integral part making Weston Turville the place it is, contributing hugely to its character and atmosphere. As a result, it is of immense value to both residents and visitors, being a much treasured unofficial recreation area, which is not only very attractive to look at but has several footpaths allowing accessibility to be enjoyed by all.

. . . There has been a fifty per cent increase in the population of the village in the last decade, and this clearly has fundamentally changed its character. The lanes and side roads are now busy thoroughfares, and the image of a peaceful village setting is more a memory rather than a present reality.

Although the development proposal brought local opposition to development in the village to a head, and while it seemed to be supported by the vast majority of the village's residents, not everyone favoured the campaign. One resident said: "People who've been here all their lives didn't get involved in the 'Save Weston Turville Action Group' because many of them have landed interests. In fact half the village wasn't talking to the other half". One farmer gave some indication of this when he said: "The 'green heart' is a place to walk the dog and tip the rubbish. There's no farming in it so it falls into disrepair and logically becomes ripe for development. The action group defending the 'green heart' are people who've moved in – NIMBYs. I think they are stupid to defend it". However, the number of farmers and landowners living in the village is small and others supported the development for different reasons. Although the majority of the letters received by the planning department in response to the application were of the character of those quoted above, several were in support of the development. One said: "I agree with the [development], as I would like one of the Low Cost Houses, as I have lived in the village all my life, and would like to carry on living there when I get married. At the moment the house prices in the village are too high for those who have lived there all their lives". Another young respondent argued: "My parents live in the village and my first home has had to be in a one bedroom cluster in Hayden Hill. It is now worth about £62,000. The cheapest house in the village has been an £80,000 semi-detached house. I am sure their are many young couples in the same position".

By the time the application came before the Planning Committee in August 1989, the planners were well aware of the level of opposition in the village. They recommended refusal, and the Committee abided by this recommendation for the following reasons:

- the development was contrary to policies in the Structure Plan, in that the development was not considered small scale in the context of a village such as Weston Turville
- the loss of open land would be detrimental to the character of the village
- it would be detrimental to the amenity of occupants adjacent to the site
- there would be an "urbanization" of existing footpaths and a subsequent loss of amenity.

The immediate response of the developers was to go to appeal. In their appeal documents they pointed to the inadequacies of the housing land-supply figures and the lack of co-operation with the development sector in their production. They argued for the need to review the supply estimates. However, as opposition mounted, the Church Board decided to withdraw the application. There seemed to be three main reasons for withdrawing it at this late stage. First, the developers had initially underestimated the strength of local opposition, which seemed to grow as the appeal came closer; more villagers enrolled in the campaign over the period. Secondly, the reasons for the District's refusal represented the groundswell of opinion within the village. The reasons themselves seemed to fully justify the stand being taken within the village. Further- more, although the NACRT was initially enrolled by the Oxford Diocesan Board as part developer of the homes precisely because it could provide legitimacy for the proposal in terms of low-cost provision, the former withdrew its support since it would not accept the priority being given to the private housing development. Joint ventures with private developers can exacerbate the problems of specific housing associations and bodies (such as NACRT) who are committed to helping low-income people find rural low-cost housing.

A final contributory and perhaps crucial factor in the withdrawal of this proposal was the onset of recession in the housing market. Two years after the development had first been proposed, conditions in the housing market looked much different from their appearance at the beginning. It may well be that the developers decided to make a strategic withdrawal in face of adverse conditions in the market and in the village.

There was a sense among many residents and eventually among the planners, that the village had reached its capacity and that any large-scale development would finally destroy its precarious rural character. Cognisant of the neighbouring development pressures between Ayles-bury, Stoke Mandeville and themselves, Weston Turville residents saw the proposal as the final "nail in the coffin" rather than an opportunity to create a new rural space. As one resident antagonistic to the proposal argued: "My husband and I moved to this village four years ago simply because we wanted to be part of a village community, to walk through quieter streets, less polluted, with less traffic. To be surrounded by fields instead of homes or shops" The development was clearly seen as a threat to these qualities of village life.

Weston Turville became to all intents and purposes a "suburban vil-lage". For instance, between 1981 and 1991 the population of the village increased by more than 50%, from 4,800 to 7,500 (1991 Census, figures at ward level). However, its distinctive identity has not been completely undermined by this influx of incomers. The "green heart" of the village became a symbol of its status as a village. Nevertheless, the class compo-sition of the village has become increasingly homogeneous. The types of housing that have been made available have been of the larger suburban sort and these have been affordable only by middle-class incomers. Thus, the villagers' "success" at keeping out further development only begins to exacerbate the broader strategic problems of housing need. Keeping villages "rural" generally means building houses that conform to the "character" of the place. These will therefore generally be "up-market" expensive houses well out of the reach of many "traditional" rural resi-dents whereas those on low incomes will find themselves increasingly excluded from the new rural spaces.

Conclusion

In this chapter we have examined three case studies illustrating different types of housing provision. What was common to the first two cases was that they allowed the middle-class access to areas from which they were previously excluded. As villages change and property becomes transformed, so the highest bidder gains access. As we have seen, this results in the exclusion of others. This sets the scene for the other development processes and future housing development. The more housing developments take place, the less chance there will be of future development in the locale. The last case showed the limits of development. In a village which has expanded extremely quickly, a halt must eventually be called to the pace of development if the character of the place is to be maintained, or it will effectively disappear as it merges with a nearby town such as Aylesbury.

The various expedients adopted by the Council to meet the needs of those who lie outside the rural housing market have been only partially successful, and will remain so while landowners and developers give priority to open-market housing developments. These problems of provision, and the dwindling amounts of land available, mean that concern is growing at the local level over access and affordability. There is a growing realization that public intervention of a more comprehensive type is needed, at least in principle. In our household survey 70% of respondents said they wished to see more council housing built. The problem for the planners, and indeed the local residents, is both how, and more specifically where, this should take place. What is also of particular significance is the difficulty in obtaining any clear measure of housing need. While, as we argued at the beginning of the chapter, housing land supply is a sharply contested process fought out through highly regulated modes of representation, when it comes to "need" the mechanisms of measurement are weakly developed. This may begin to change as SERPLAN and planning authorities begin to allocate more potential sites for affordable homes as part of their land allocation exercises. Nevertheless, shortages of low-cost rural housing have placed additional burdens on the existing towns of Aylesbury and Buckingham.

In summary then, modes of housing regulation incorporate particular conceptions of class and space, which are juxtaposed around formal systems of allocation and provision. This is enmeshed within a protectionist planning policy which ensures that competition for housing stock is fierce and those with the most resources (income, wealth, mobility, etc.) naturally "win out". Thus, rural settlements come to be increasingly dominated by a rather narrow social strata. However, once established in a place, these strata are likely to go to great lengths to defend the neighbourhood against further development. In the latter two case studies the

Conclusion

development proposals were opposed by existing residents. Thus, an entrenched middle class will fight to exclude others from the village and in the process will deprive other potential middle-class residents of their place in the country. Moreover, the poor and many traditional rural residents find themselves squeezed out of ever more exclusive villages.

CHAPTER 4

Developing places:
new villages for old

Introduction

In the preceding chapter we focused on the housing land development process and on how this is conditioned by a range of actors operating in different contexts and at various regulatory scales. We now turn to examine how this development process affects places in different ways at different times and illustrate some of the social consequences as outcomes coalesce, putting places on particular development "trajectories". In this chapter we want to examine a range of these trajectories by focusing upon three villages in Aylesbury Vale. These have been chosen to emphasize the variation existing in the locality and to allow the examination of the local social impacts of land development. As we have discussed elsewhere (Marsden et al. 1993 Ch.6), we see localities as "meeting places" where various development networks intersect, leading to a range of outcomes which in turn provide the basis for future action. As we argued in the previous chapter, housing developments are the outcome of both local and strategic processes, being influenced by the competing representations of local and strategic actors. Furthermore, housing development allows new social groups access to rural space often resulting in action to constrain further development. It is therefore necessary to examine more closely the processes of land development at the village level and the social change that surrounds these processes.

Through in-depth interviews with key actors we have attempted to piece together selected social profiles of three villages, related particularly to the combined effects of land development processes occurring in each. Of particular importance in this discussion is the relevance of *place* to village residents. Our survey evidence suggests that village attachment and concern for the village environment, rather than being diminished by the high levels of social and spatial mobility, are actually enhanced. Contemporary village residents are now keen to reconstitute space in their own terms and in ways which reflect their aspirations and livelihoods (see Marsden et al. 1992). High levels of mobility in residence or work do not diminish an individual's commitment to the relevance and distinctiveness of a rural place. Thus, the provision and availability of rural facilities and amenities, including access to open countryside, becomes particularly significant in places like rural Buckinghamshire where residential and occupational mobility are relatively high. The social experience of movement can thus increase attachment to place. Such attachment can then lead residents to put particular demands on the planning system to deliver environmental goods, partly by restricting the amount of development. While there are similarities in the survey data in the patterns of mobility between urban and rural households, and considerable urban experience amongst many villagers, it is clear that village residents demand significantly distinctive and "authentic" rural environments which need to be upheld in the face of development pressure. It is as "outcomes" of these sorts of processes that we can assess our three villages.

The villages we have chosen to investigate in detail are: *Swanbourne*, an estate village in the north of the district. It is dominated by an estate landlord and the collection of tenant farms, found in and around the village, has maintained the agricultural environment by keeping development pressure at bay. It is, on the surface at least, the quintessential paternalistic rural village; *Weston Turville*, which, as we noted in the previous chapter has experienced profound changes. Situated in a "green wedge" of land between Aylesbury and the market town of Wendover, it has during the past twenty years experienced rapid in-migration. It is very much a large, suburban "metropolitan village" which has almost lost its distinct rural identity. Agricultural activities have been marginalized and the central village green (glebe land) remains the major characteristic of the traditional village environment, and a focus for concerns surrounding the perceived over-development of the village; and *Wingrave*, lying to the east of Aylesbury, which has largely lost its agricultural *raison d'être*. It has been subject to successive waves of middle-class in-migrants and different types of land development have occurred over recent years. Despite this the village is socially mixed and has a variety of building styles. The landed estate was sold off to farmers long ago and the village has experienced new forms of class colonization.

These villages are not just distinct; they represent the variety of different rural spaces that may be found within one locality. They thus allow us to understand how local action is embedded in different contexts. The regulation of these contexts by different actors and institutions is crucial to their maintenance or to the degree of change they are likely to experience. For instance, will Swanbourne and Wingrave become more like Weston Turville? Or do Swanbourne and Wingrave owe their very character to the fact that Weston Turville has changed so radically? At this micro level we can begin to track some of the different social consequences of these developments and to see how the social and the spatial intersect.

The village case studies

Village study 1: rurality retained

Swanbourne is situated two miles (3 km) east of Winslow and nine miles (13 km) north of Aylesbury in the heart of Aylesbury Vale. It is midway between the two growth poles in the area – Milton Keynes and Aylesbury. With a population of 400, which has virtually remained static for the past twenty years, the village sits in the most "rural" part of the Vale and is itself still a truly rural village with extensive parts of the village fabric remaining unchanged since the 17th century: "The existing loose knit pattern of development also reflects the early medieval field system, the remnants of which can be seen in the number of narrow sunken lanes, high banks and paddock boundaries as defined by lines of remaining hedgerow trees" (AVDC 1989b). The village retains its rural character largely as a result of the green space still within the village boundaries. In fact the division between the village itself and the surrounding countryside is much less evident that in either of our other two villages: "The surrounding countryside, although now missing its distinctive elms, is richly agricultural and provides a luxuriously treed parkland setting, which in places permeates through to the heart of the village" (AVDC 1989b).

It is an estate village. The estate itself is still intact and passed into the hands of its current owners in the mid-19th century, when Swanbourne House was built for the family (it is now a private preparatory school). It is the estate, and the social and economic conditions associated with it, which have allowed Swanbourne to largely retain its traditional character. It owns all but 20 ha of the land within the parish and half of the properties in the village. According to one respondent, who has lived in the village for over ten years, the social structure can be divided three ways: first, the estate landlord and his nine tenants; secondly, freeholders living on land previously owned by the estate; and thirdly, the council tenants. The grandparents of the present owner built a number of Victorian houses in the village and three pairs of cottages between the wars. In the 1960s and 1970s investment went into modernizing the old cottages. There have been only minor developments since then, with five houses built in the past ten years. Consequently, the fabric of the village is still very much constituted by the working farmsteads of which there are seven within the village boundaries. The policy of the estate farm is currently to take the tenanted farms "in-hand" as they become available; in the past ten years the size of the estate farm has increased from 270 ha to 350 ha. The tenanted farms range from small holdings of 8 ha to the largest holding of 150 ha (105 ha of which is rented from the estate). There is some arable farming on this land (around 80 ha) but the rest is grassland given over to livestock. Neither is this intensive livestock farming; it comprises sheep, store beef and suckler cows. The pattern of farming on these tenanted farms could be characterized as extensive and traditional. Only on the farm not owned by the estate is the whole area (47 ha) given over to arable farming, but this too is on a small scale. Consequently, incomes are low (financial information for one medium-sized farm to which we were given access calculated annual net farm income to be just £3,390).

The main influence on development in the parish can be discerned by briefly considering the strategies of two actors – the landlord and a tenant.

The landlord: "keeping the assets" The landowner, now in his mid-sixties, started managing the estate when he left the Navy in 1966. The estate has been in the family for 150 years. The farm was a relatively small holding of 65 ha, with mainly livestock enterprises, with the rest of the estate let to sitting tenants. The landlord was not satisfied with the way the estate was being run. On taking over the farm he began to take land "in hand" when it became available, both for the purposes of increasing the size economies of the estate farm and to reduce fiscal commitments and tenant responsibilities. He diversified the enterprise (to sheep, beef and arable) on the advice of the Milk Marketing Board and the Agricultural Advisory Service, using capital grant schemes for drainage and buildings improvements.

His son was made a partner in 1989, followed by his wife, both for tax reasons. At the time of interview there were seven employees: a farm manager, a secretary and five agricultural workers. This labour requirement has increased as farms have been taken in hand. There were originally (in 1966) only two full-time workers. The landlord works a limited amount of time on the estate (about one hour a day). The estate farm covers 350 ha with 182 ha of cultivable arable land and 168 ha to permanent pasture.

He argued during interview that non-agricultural diversification was not really considered as a serious strategy on the estate farm since there was little financial need. Nevertheless, in 1989 he sought permission for a barn conversion on a redundant building. He saw housing (tenants) policy, land tenancy laws and the Employment Protection Act as reducing his powers as a landlord. He wanted more freedom in the private rented sector in order to control new lettings and to relax farm tenancy laws, thus encouraging fixed-term lets under his control. The maintenance of the private ownership of the land base was seen as critical to the village, and he was expecting his only son to succeed to the farm.

With an annual turnover of £250,000 and a combination of dairy livestock and arable enterprises the overall strategy by the early 1990s was to "keep the assets" and to minimize risks and debts. He saw the estate farm increasingly in uncertain terms, arguing: "It is uncertainty which is the major job; more than anything else falling prices and falling income. How long will subsidies last? Who knows what the future will hold?"

97

Faced with falls in prices, he said he would attempt to increase the quantity of production, also arguing that he would consider selling land for development rather than attempting to diversify. He had purchased £30,000 worth of milk quota in order to sustain the enterprise in 1990. He intends to pass over the business to his son in five years.

The tenant: "converting the assets" Unlike many of the villages in the study area, and partly as a consequence of the persistence of the landlord-tenant system in Swanbourne, the central part of the village is still dominated by tenanted farms. Mr C. rents 65 ha from the landlord as well as owning 38 ha in Wales.

His father had previously rented from the landlord and, when Mr C. was 30, he became a partner with his father. Two years later the tenancy was transferred to him. By the late 1980s most of the land on the Swanbourne farm was down to permanent grass with sheep rearing being the main enterprise. Bouts of arthritis had forced the farmer to relinquish the sheep, beef and dairy enterprises although the compensatory payments in the early 1970s were helpful for the withdrawal of dairy cattle. In 1984 his mother died leaving him a bungalow which he sold, using the equity to purchase the sheep grazing farm in Pembrokeshire. The farmhouse was sold in order to pay off the £10,000 per annum interest on the mortgage for the Pembrokeshire farm. During the survey period (1987–91) Mr C. was thus in the process of circulating equity for debts and investment as he redirected surplus capital towards the Pembrokeshire farm. As an established tenant his fixed capital costs were lower than for owner-occupiers, and the investment in the Pembrokeshire land was seen as a longer term strategy. By 1990 the farmer had rid himself of all mortgages, long-term loans and overdrafts. He argued: "Most of the neighbouring tenant farmers have chucked the towel in. I'm proud to have survived. I take up any government schemes whenever I get the opportunity. I am looking forward to being paid not to keep sheep!". His wife and family had left him in 1985 and since that time the farmhouse had been let to lodgers. With some land re-let for horses and a shed let to a car restorer, Mr C. had diversified his income sources.

Nevertheless, he was pessimistic about the future. Central government and the European Community were seen as unsympathetic to the accumulated value farmers had bestowed on the land, and he perceived that there was little opportunity for young people in farming generally and in tenant farming in particular. Landlords taking land "in hand" was a symptom of these uncertainties. He argued: "It's a dead loss to rent land". He hoped eventually to obtain planning permission for a field centre on the Welsh farm and then sell it for profit. He argued for the break-up of the large estates and for more government support for the small family farm sector. With his daughter only helping occasionally (with the horses) on the farm there was no aspiration towards succession. Mr C. was eventually hoping to realize his capital assets once his tenancy was relinquished. Moreover, he was hoping to "second guess" policy changes by overstocking sheep so as to maximize compensation when it arrived. Premiums on ewes were encouraging stocking densities and, with the ownership of the Welsh farm, sheep could be moved to Swanbourne for fattening. The main reason for these strategies concerned the need to clear debts, to pay interest and "to raise cash for the landowner and for my own future use". In 1990, he said that he received £15,000 per year from the farming plus £1,000 from other part-time work and of this £6,000 was paid in rent and overheads.

We can see from these cases that the gradual decline in farm incomes and the need to constantly appropriate value from owned and tenanted land sets an important context for the relationship between landowner and tenant. While they obviously hold structurally distinct positions in village life and although both are key figures in the social and political life of the community, they are also forced to play different rôles in the re-commodification of land rights. The landlord, while continuing to espouse his paternalistic village responsibilities, is forced to take farms "in hand" and to convert barns into houses. In the future he contemplates selling portions of land. The tenant recycles inherited family wealth beyond the locality, seeing this as a process of asset realization. While the landlord-tenant relationship still plays a key rôle in the maintenance of the community of Swanbourne, it is not immune from the retreat of agricultural "productivism" or the prospect of potential development gains from persuading the planning authorities to agree to the selective release and conversion of agricultural land.

The scope for the latter can be ascertained from a review of AVDC's Planning Bulletin which between 1987 and January 1991 reveals applications entered for the building of five houses (one development of three and one development of two), an application for land as a builders' yard and light industrial use of farm buildings (see Fig, 4.1). There was also an application in 1990 to open a landfill site in an existing small quarry on tenanted land owned by the estate. The proposal intended the site to be used for builders' wastes, deposited over a period of five years, finished with a layer of top soil and reseeded. The landlord supported the scheme and the Parish Council had no objection. The District Councillor (also a tenant farmer in the village) recommended refusal and the District Council voted to oppose the scheme on the following grounds: (a) the landfill operation would have a seriously detrimental effect upon the visual amenities of the area; (b) the roads leading to the site are inadequate to deal with the intensity and character of the traffic which would be generated by the development and this would be detrimental to residential

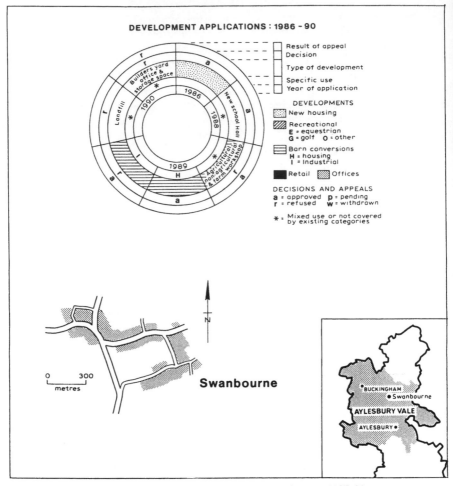

Figure 4.1   Development pressure, Swanbourne 1987–91.

amenity in surrounding villages; (c) the proposal would have adverse effects on the ecology of the area; (d) there is no need for this facility (AVDC Planning File). The County Council subsequently refused permission because "the length of time proposed for the operation would not result in an improvement of the land for agriculture within the shortest practicable time". Furthermore, the County Engineer felt the highway was not adequate.

The unsuitability of the roads in Swanbourne for most heavy transport activities is a deterrent to development. This was the issue in another case where a farmer put in an application for non- agricultural storage and a farm workshop in his farm buildings. The farmer was using his buildings as a farm machinery workshop and as a storage facility when an enforcement notice was served upon him after the planning authority received reports that this type of activity was taking place on the farm. Consequently he applied for planning permission in January 1989. There were no objections to the proposal from the Parish Council, although there were seven letters of objection from local residents, mostly on the grounds of increased traffic on the small single track lane which serves the farm and runs alongside the nearby houses. The farm is situated in Nearton End, an area which originally comprised a grouping of farms, farm buildings and a few cottages. More recently there has been some infilling but the four farms in the "End" still dominate and the planning officer's report described it as "rural and very quiet in character". The applicant's farm is in a cul-de-sac at the end of the lane and the planning officer's report concluded that: "its location at the end of the cul-de-sac which will involve increased heavy traffic on the narrow lanes is likely to cause a detrimental effect on the amenities of the occupants of properties in Nearton End by virtue of noise and disturbance". In the light of this it was recommended that permission be refused for the following reasons: "(i) The roads serving the site are inadequate . . . (ii) The use has, by the noise and disturbance it creates, and the type and volume of traffic generated, produced a significant loss of amenity for the occupants of Nearton End and of neighbouring houses in general . . . (iii) The continued use detracts from the character of the area, which is under consideration for designation as a Conservation Area" (AVDC Planning File). Permission was refused in May 1989.

The applicant subsequently appealed against the decision. The agents operating on his behalf put forward a similar set of arguments to those made at the original application, i.e. that the farm alone could not yield sufficient income for its occupier and some form of diversification, in line with government policy, was therefore necessary. Furthermore: "the roads serving the farm are used by large articulated vehicles not only to collect harvested grain from the farm but also to deliver materials in bulk such as seeds and fertilizers. The vehicles used in connection with non-agricultural storage are smaller than those employed in connection with agricultural uses". This time letters of support were received by the planning office, significantly one from the estate landlord and others who had used the farm machinery side of the business. The County Engineer, however, countered the above comment from the agents in his evidence to the appeal:

"It is accepted that other large agricultural vehicles already use the road to service the existing farms, however, Nearton End is within a farming community and consequently such vehicles must be tolerated, despite the problems which they can cause . . . . The appeal site buildings do not generate any agricultural traffic other than when used for the storage of grain harvested from the appellants land and, therefore, their re-use for storage and repair work must result in additional large vehicle usage of the unclassified road. The highway authority is of the opinion that the unclassified road is inadequate to accommodate this additional traffic". (AVDC Planning File)

The Planning Authority took the view that an agriculturally related workshop would be acceptable but resisted the proposed storage of non-agricultural items in the building. This was upheld by the Inspector who argued that: "I agree with the council that a farm workshop is an appropriate activity in a rural location . . . . I agree with them that this should not involve the repair of vehicles, for the same reason as that which causes me to find the commercial storage element to be objectionable, that is, the generation of further traffic". So despite the fact that the non-agricultural activity was unlikely to provide any more disturbance than agricultural activities it was concluded that the former was inappropriate in Swanbourne. Within a farming community a certain amount of "disturbance" must be "tolerated". Non-farming activities of a similar kind are, however, viewed as "objectionable".

We have examined this case in some detail in order to exemplify the constraints preventing farmers from diversifying their holdings in any significant fashion. Primarily, the physical infrastructure of the village militates against any significant development involving increased traffic, given the priority placed on landscape considerations. The village has now been designated a Conservation Area. Such a status is given to "areas which [the local council] consider to be of special architectural or historic interest, the character or appearance of which it is desirable to preserve or enhance" (AVDC 1991a: para. 5.14). Furthermore: "In dealing with applications for planning permission on land in or adjacent to a conservation area, the Council will have regard to the effect the development would have on the special character of that area and the higher standards of design within it . . . the council will wish to satisfy itself as to whether the proposal would preserve or enhance the character or appearance of the conservation area" (AVDC 1991a: para. 5.23). Any subsequent development will have to adhere to the restrictions that this imposes.

Although adding greatly to the rural character of the village, the farms do not account for a great deal of the land within the village boundaries. The estate owns roughly half of the properties, the rest, with the exception of the council houses, are freehold. The rôle of the landlord is, therefore, crucial to the future development of the village. He perhaps recognized this when he said: "It's fair to say that because I'm so centrally positioned, what my attitude is perhaps is more important than that of the planners". The village retains its rurality through the structure of the estate and the maintenance of the tenanted farms. The planners are aware of this and for the foreseeable future these farms look set to struggle on, severely restricted in their opportunities to convert their buildings or diversify their activities. The attitude of the landlord to development is that: "I would like to maintain a modest increase in the village population to keep the amenities without swamping the village". Gradual development seems to be his aim and without his restraining hand the landlord felt that the village would be subject to more development pressure: "Without constraints there would be a lot of infilling, new developments and unsuitable development". However, further constraints are imposed by the planning system. The rôle of planning is summed up in this comment by a planning control officer: "There are few applications, so there is little development control". The designation of Conservation Area status would seem to add more formal weight to planning restraint. According to the planning officer there is a "strong conservation argument in Swanbourne the longer it remains unaltered". This does not mean that there is no scope for development, however. Another planning officer argued that: "There are sacrosanct areas of Swanbourne but other areas where infill may be allowed. Some places allowed the integration of the 'gap' and the property [but] the rural character would be lost if the gaps were lost". This attitude depends to some extent upon the overall context in which plans are being formulated, as the former planning officer intimated when he said: "There are targets for houses in the Structure Plan and we have commitments to meet those targets. We are not looking to achieve substantial numbers in the villages so we can have a strong conservationist bias"

All these factors seem to indicate that the village will retain its present shape into the foreseeable future. The location of the village, midway between the two growth centres of Milton Keynes and Aylesbury, means that the pressures on the village may be kept at bay but are likely to remain lurking in the background. Certainly the villagers are aware of this, as one said: "People are very frightened that they are going to be swept away by Milton Keynes". At present however, the main impact of Milton Keynes has been on the levels of traffic through the village, an issue mentioned by most of our interviewees. It is made more noticeable because, as one respondent put it: "The main road through Swanbourne divides the village into two parts, it's a very busy road. Nearton End is on one side, Smithfield End on the other. This is increasingly important as both 'Ends' are little 'backwaters'. More local people live in Smithfield End whereas Nearton End is a little bit more 'nouveau riche'"

The village has been characterized as a "stable" community which has gone some way to incorporating limited numbers of incomers. As the same respondent put it: "Old Swanbourners know they must accommo-date the new to survive. Old suspicion has to be lost – there is a real sense of Swanbourne reconstituting itself". This "reconstitution" is undoubted-ly in its early stages and is likely to be gradual. One local, in his eighties and resident in the village all his life, said: "I'd like to see the village develop but I want it to keep its character and remain the kind of commu-nity I'm used to. But there's an ageing population and that's a bit of a problem". However, a planning officer argued "people in the village are in a bit of a pickle because they are tenants so they are stuck there. It's a very stable community although not as ageing as you might think, there are a few young families". While it may be true that there are a few young families the perception remains, amongst those who live in the village, that it is "old" Furthermore, the stability and the overall age structure seem to condition the existence of the "community". The District Coun-cillor, also Chair of the Parish Council, said: "It's difficult to get the old indigenous population to be active in the community – by the very fact they have always been here they tend to be less dynamic. We are grateful for newcomers. We have an old retired, ageing population". This was echoed by another resident who had lived in the village all his life: "The character of the village has stayed quite the same due to the [landlord]. This has depressed the character of the village and led to apathy; they rely on [the landlord]. Organization is difficult because nobody bothers and people who've moved in keep themselves to themselves"

Nevertheless, there are village societies, most notably a Friends and Neighbours Club, which organize a variety of activities mainly for the older residents. There are regular whist drives and occasional barn danc-es. What does seem to be lacking is any activities for young people. As the District Councillor said: "Occasionally a youth club rears up and then goes again. There's no cricket club and no football club; there are sporadic attempts to provide these but they usually come to nothing". The more formal village institutions still function but, again, an air of apathy seems to surround them. The District Councillor said: "The Parish Council is made up of people who've been here a long time . . . It deals with things like footpath lighting, maintenance of village assets, village hall and play-ing fields. There aren't really any burning issues or conflicts on the Parish Council. It's hard to get people to take much of an interest"

There are two churches – one Methodist and the other Anglican. The Anglican church has a congregation of around 40 and is run in conjunc-tion with three other parishes. On certain occasions the three parishes come together to worship, at Easter or Christmas for instance, but the villagers tend to prefer their own services. According to the local vicar, the Church Council is made up of "an absolute cross section". However, there still seems to be a close relationship between the church and the landlord's family; as the vicar put it: "That he [the landlord] is a patron of

104

the church gives a feel for the relationship between the church and the family". This kind of relationship still exists between the landlord and the local school. A teacher said: "[the landlord] is the benefactor of the school – if we want a gate made he gets it made, as a governor he has the welfare of the school at heart. He doesn't like the numbers so low, otherwise he never interferes. There are people who think he owns the school but they have an old fashioned view of the squire". The school itself is in decline, with numbers having fallen from 59 in 1980 to 22 in 1991. It has one full-time teacher and another part-time. A small proportion of the pupils come from outside the village (attracted, the teacher told us, by "a small school, no bullying, teacher attention, open access"), and there is an even spread of those from within the village derived from the various social groups. The school still feels it gives a "rural" education; as the teacher put it "we use the farms around here, the church and the churchyard etc., for teaching purposes. We use the first hand experience of the environment, we go to the farm next door to see calving". The future of the school is precarious however, particularly with the introduction of Local Management of Schools (LMS), which bases funds upon numbers. As the teacher said: "The school here is in danger and LMS has made that more stark. Now the Education Authority may say we cannot carry you in the expectation that something may happen when it may not" So the village school may close. The private preparatory school seems to be in no danger, however. It now has 300 pupils (it started life with 100) and has begun to take in children from the age of three. This school has become a major source of employment for women in the village.

In general then, Swanbourne betrays a great sense of stability and continuity and this marks it out from many other villages in the locality. Although there has been some degree of in-migration the village retains a very mixed, yet traditional, social complexion. It has council houses in the village centre, ageing agricultural tenants who are "muddling through" on low incomes, and a reasonably high proportion of indigenous inhabitants. The landowning and agricultural forms of paternalism cast a long shadow across the village, reinforcing the pattern of low levels of development which have protected the traditional social fabric. However, the result has also been an ageing population and a degree of inertia in the community. This provides a useful contrast with the following two villages.

Village study 2: the eclipse of the rural

Weston Turville is a large village situated two miles (3 km) southeast of Aylesbury and just beyond the Chilterns AONB. As we discussed in the previous chapter, the location of the village means that it has been subject to pronounced development pressure. It has grown enormously over the past two decades, particularly during the housing boom of the 1980s. The result is: "A sprawling village . . . . a curious mixture of an old world charm and modern suburbia. The casual motorist may be forgiven for assuming that Weston Turville is a modern settlement. This would, however, be a gross misconception for within the built fabric of the village are discreetly identified areas of special 'townscape' the value of which it is important to protect" (AVDC 1991b). This mixture of "old world charm" and "modern suburbia" has come about as a traditional Buckinghamshire rural settlement has tried to accommodate a significant number of new housing developments. The village emerged from a number of "endships" and the form of the village can be ascribed to "frontage" development along the roads linking these "endships". Between these roads infilling has occurred and there has also been the development of short cul-de-sacs, mainly during the 1970s and 1980s. Earlier rounds of this occurred during the 1950s under the aegis of the former Aylesbury Rural District Council, and were taken further by the District Council during the 1960s and the early 1970s when permissions were granted for large new estates and a significant amount of infill. The bulk of the permissions were granted in response to the Secretary of State for the Environment's call for the release of more land for housing at that time.

The pattern of development around the roads linking the "endships" resulted in the village centre, composed of three fields, being virtually surrounded by housing developments. The final piece of surrounding came with an application from the Church (the Oxford Diocesan Board) in the mid-1980s to build eight houses on a glebe land site. Although this development was vigorously opposed by the District Council it was eventually passed on appeal and effectively surrounded the "green heart" of the village leaving it vulnerable to further development applications. As we have already discussed in Chapter 3, the Oxford Diocesan Board then applied for considerably more development on the glebe land in 1989. This proposal brought opposition to development within the village to a head and resulted in the formation of a "Save Weston Turville" Action Group.

Although, it was, and is, AVDC's policy to allow settlements such as Weston Turville to grow and evolve at a rate which does not fundamentally alter their character, Weston Turville has expanded to such an extent that a substantial surburbanization of the village has taken place. The construction of one site (58 houses) "opened the door" for subsequent development, allowing a pronounced change in the social and material shape of the village to occur. As a result, unlike the cases of Swanbourne and Wingrave, it has subsequently been harder for the council to argue, in resisting development, that new proposals now go against the "character" of the village.

The earliest of these developments were not welcomed by village residents. As one long serving member of the Parish Council said: "Most of the early developments were opposed by the local community. Infilling is okay and was acceptable but with large developments we get too much construction and too many cul-de-sacs". The scale of the changes to the village fabric meant the community was affected in ways difficult for "locals" to ignore. As another long-term resident said, this "brought in a lot of young people, young families . . . We used to know everybody in the village but after this development we didn't". With the subsequent growth of the village the social character changed further. Another long-term resident (17 years) now living elsewhere, said: "There was a very close knit community when we arrived in the village – very mixed; farmers, artisans, professional people, a well balanced mix. You had to live in the village for a few years to be accepted. Those days have gone; it's now slightly 'yuppie-ish'. Up to six years ago it still retained its 'rural feel'. We left because it was losing its quiet sleepy backwater character"

This pattern of development continued into the late 1980s (see Fig. 4.2), bringing social and material changes which heightened the feeling of long-term residents that Weston Turville was being forced to cope with too much development and social change. Another member of the Parish Council said: "Gradual development can be absorbed by the village and the community here. But there have been large developments. What you tend to find is that the people who move into the big developments only tend to stay here for a few years, on average about seven, before moving on somewhere else. The development proposals coming forward in recent years have been large scale; you don't get people applying for planning permission for a house for a son or a daughter but for five houses for outsiders". The influx of population, and the provision of housing for more incomers, has made it difficult for local people to find suitable (low-cost) housing in the village. A long serving member of the Parish Council, a "local", said: "You could say that I used to know everyone in the village but not now. It's not possible to know everyone now because so many people have moved in, but also many young people have left. Some youngsters would like to stay in the village but it's not possible. For instance, some five bedroom houses have been built . . . No one in the village can afford them".

Many of the farms situated within the village envelope have been turned over to housing and there is now only one working farm left. Nevertheless there are still "local" long-term residents living in the village. According to a resident born in the village "forty or fifty families now are the 'core' of the old village who are the only families to have been here for more than 15 years". These residents are still active in the village institutions. The Chairman of the Parish Council said: "I've lived here for 35 years . . . I'm Chair of the Parish Council, also on the Committee for the new village hall, also Chairman of the School Governors. I've been on the Parish Council for 20 years".

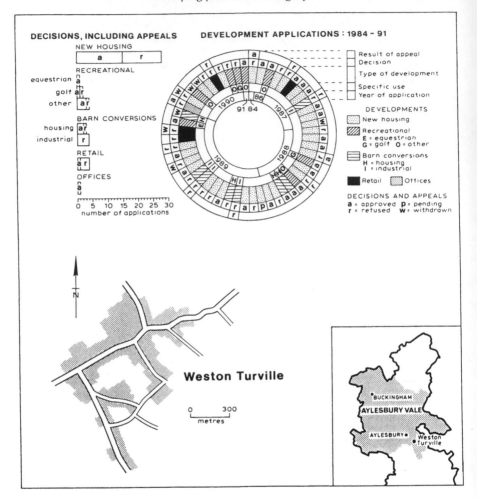

Figure 4.2 Development pressure, Weston Turville, 1987–91.

Weston Turville has a number of thriving community institutions which have gone some way to accommodating and incorporating the incoming population. The incoming population has not, however, taken over these institutions, as the above response indicates. According to one long-term resident: "There are within Weston Turville lots of little communities around certain institutions but now there are also lots of home-centred people". Where incomers have wanted to participate it seems that they have not had too much difficulty. The Parish Council, for instance, is evenly split between newcomers and locals but the Chairman and Vice-

Chairman are long established residents. The Parish Council plays a crucial rôle in keeping the village informed about planning applications and development. According to the Vice-Chairman who also oversees the planning side of the Council's work:

> The Parish Council gets to see all the planning applications for the parish. We oppose about 10% of them but the bulk are for things like garages, conservatories, etc. There was a big change when AVDC started posting out information on planning applications (before that we had to go to the Council Officers and ask to see them) then the planning subcommittee started informing people if there was a development proposed next to them so they were aware of what was happening. There are always objections to large-scale development and the Parish Council always objects.

The Parish Council recently lost its last farmer member and, unlike Swanbourne's, retains no real relationship to the farming community. It is effectively a village organization concerned with the problems of managing growth and relatively large-scale housing development pressure.

There are a variety of social service organizations in the village, engaged in such activities as collecting prescriptions, bed-making at a local hospital, meals on wheels, and a visiting service. These services are primarily run by women. Other women's organizations are the Women's Institute, with around 30 members mostly drawn from older villagers; the Monday Club, with members of all ages who meet to play bingo, etc.; and the Toddlers Group, for mothers and young children. The view was expressed by certain respondents that it is the women of the village who really keep the village institutions going and make Weston Turville into something resembling a community: "You don't hear women saying there's nobody to talk to in the village" said one long-term female resident. Many men, especially those commuting to work elsewhere, tend to be away from the village for significant periods. In the village their activities seemed to be mainly dominated by sport. A majority of men sit on the Amenity Committee, and there are golf and squash, sailing, sports and fishing clubs. Men are also active in the Horticultural Society. Despite the number of active clubs and societies they still fail to add up to a thriving coherent "community". As one woman who has lived in the village for 20 years put it: "There are a lot more clubs but the community spirit seems somehow lacking. When I used to walk up the lane everyone would wave and speak, but now no one speaks. Everyone walked and talked, now everyone moves in a car". The individual home-centredness of many of the new residents means there are very many people in the village who simply do not participate in the various village activities.

The school has however been markedly altered by the growth of the village. In 1970 it moved to new premises and at that time held 163 pupils but by the time of our study (1991/92), it had 257. Staff include the head teacher plus eight full-time and four part-time teachers. The rise in numbers is partly accounted for by the influx of people to the village but also by the school's good reputation, built up during the past 10 years, so that parents are now sending their children out of Aylesbury to Weston Turville. This may also be linked to the desire of some parents to have their children educated in a rural area where, according to the head teacher, "the children have less of the 'pseudo-sophistication' that they have in urban areas". Furthermore, the type of people who are moving in are, according to the same respondent, "professional people . . . with high aspirations for their children and if they are not happy they'll send their children to private school": This puts pressure on the school to meet the requirements of such parents and the sharp increase in numbers indicates that it is doing so.

The church, as in many rural villages, has a small congregation of around 30, but remains a focus of village life. However, the Church authority, the Oxford Diocesan Board, recently upset many in the village by their attempt to develop the glebe land. The Church owns part of the village "heart" and has already developed on the edge of the site where eight houses have been built. As we saw in the previous chapter, in 1989 the Board put forward a planning application for a comprehensive development on the glebe land which included affordable housing. Although this proposal was opposed by a majority of residents, some "local" people, including farmers and landowners and others seeking affordable housing, supported the development. However, their numbers are small. The land outside the village is part agricultural and part recreational. There is only one owner-occupied farm of 210 acres. The other major land holders are an agricultural college, a golf course and a sports club. Unlike Swanbourne, the rest of the land, both inside and outside the village, is fragmented into small fields, many owned by individuals. We have managed to identify eight landowners of these "fragments", some of whom live in the village, some live elsewhere. They stand to gain from the development of their parcel of land and may therefore hold different views of the development process in the village than the newer residents. The problem of low-cost housing provision is also one which concerns some villagers in Weston Turville, mostly consisting of young locals who wish to remain in their place of birth. The package offered by the Oxford Diocesan Board was clearly tailored to appeal to those struggling to remain in the village. As the parish's District Councillor commented: "You can argue that we should have starter homes, but land values near Aylesbury are very high. Landowners in this area will simply hold out to try and get the maximum development value for their land". This "local" support was overwhelmed by the scale of the opposition to the development. The opposition was initially mobilized by residents living in the houses built on the glebe land, adjacent to the proposed development. Clearly they would be severely affected and would no longer be living adjacent to a patch of "green space". However, these were also the newest arrivals in the village and, as such, could be accused of "pulling up the drawbridge behind them".

Whether the initial decision to reject this particular proposal will indeed mark a turning point in the inexorable development of the village remains to be seen. The District Councillor said: "I think Weston Turville has reached its peak and now Wendover [a nearby market town] will have to take its fair share. There are potential pockets of development still, but not of great significance. There are green boundaries around the village marking the boundary limits . . . My burning desire is that the villages [around Aylesbury and Weston Turville] should not join up"

Parts of Weston Turville have recently been designated by the local authority as Conservation Areas, partly as a result of the controversy over the glebe development, so the restrictions on development are likely to be tightened. But perhaps the most important issue is that of preserving a "green belt" between Aylesbury and the villages to the south. This is regarded as crucial, given, as we saw in the previous two chapters, that this area is under immense development pressure as densities in Aylesbury increase. A recent public inquiry into the siting of a superstore on the edge of this area came out against the development on the grounds of preserving the green belt. However, there is, at the time of writing, talk of a southern bypass around the town which would run through the green belt. A planning officer, when asked about the implications of this, said: "If the southern bypass is going to go through the green belt between Aylesbury and the southern villages, can we justify resisting pressure and development in those villages if they are going to be less distinct? The overall context for development pressure is going to be important here because the Authority may be forced to concede development in this southern area."

At present the *Rural Areas Local Plan* does not anticipate any further large-scale development in Weston Turville and the small town of Wendover, lying just south of the village, is regarded as having some room for expansion. Clearly the villagers will attempt to resist any further encroachment on the green areas within the village boundaries. It is the weakening of the green belt surrounding the village which must be perceived as the biggest threat to the distinctive identity of Weston Turville. Hence, there is pressure within and beyond the boundaries of the village. For the moment no change in policy is anticipated, but in the longer term Weston Turville's position on the edge of the Chilterns would seem to ensure continuing development pressure; it is this broad context which may in the end determine the destruction of the "village" and lead to its complete incorporation into an expanding Aylesbury town.

Village study 3: development and community conflict
We can see from our analyses of both Swanbourne and Weston Turville that class formation is an uneven process with the presence of the middle class highly variable over space. The two villages provide very distinct contexts for the assertion of middle-class values and aspirations. Formation is linked both to length of residence, as the survey evidence presented in Chapter 2 indicated, and the pre-existing condition of the village. The extent of middle-class strength in particular places often becomes clear when new developments are proposed, with local – newcomer representations competing for dominance. In both Swanbourne and Weston Turville we have outlined how different entrenched "class fractions" (landlord and tenants in Swanbourne and middle-class groups in Weston Turville) maintain their hold on these spaces. We have already seen in the previous chapter how the recent conversion of barns at Wingrave allowed members of upper income groups to enter the "heart" of the village. As we shall see in this section, the social impact of new barn conversions in Wingrave, or housing estates in Weston Turville, occurs as middle-class residents gradually try to mould the identity of the village to one which suits them. This may happen over a long period, but, as we have seen, middle-class occupation of these places goes back some considerable time. Other social groups may find a reflection of themselves in these village identities while some may feel excluded.

Wingrave is a Chilterns village 50 miles (80 km) northwest of London and 5 miles (8 km) northeast of Aylesbury, lying just off the main Aylesbury–Leighton Buzzard road (A 418). The village is perched on a hilltop with land falling away on three sides. The village has a long history stretching back to the 10th century and contains features which reflect its past – village green, pond, manor house (now owned by the MacIntyre Trust for disadvantaged children), recreation ground, etc. The old "heart" of Wingrave, as this area of the village is called, was given cohesion by seven farmsteads. These farmsteads were in close proximity to each other with their lands falling away from the village.

Until 1961 the population stood at 832 and there had been little change, the only exception being the provision of a dozen council houses outside the old village. The council houses, as was common elsewhere, were differentiated from the village proper. The old village remained much the same, still with seven working farms. In 1967 a village plan was drawn up by the local council. Parish councils at this time were keen for their villages to have the "prestige" of a formal plan (reflecting a kind of sixties "modernist" approach to countryside development). Growth was seen as a desirable goal. The Wingrave plan identified room for expansion to the northwest of the village heart. It proposed the siting of a new housing estate (60 houses) between the council houses and the old village. The school, previously near the village green, would now be moved to the centre of the new estate. The plan anticipated that the population of Wingrave would double. A new school was built in 1974 with the estate then built in two phases, the second of which was completed in the late 1980s.

The next stage of development occurred in the 1980s and affected the old heart of the village. Here there was room for infilling, which proceeded at a slow but steady pace. Most significant however, was the gradual conversion of the farmsteads to small housing complexes, ranging in size from 3 to 17 houses. There are currently only two working farms remaining in the village; the rest have all been converted to housing. In total, infilling and conversions have led to approximately 60 new housing units and the population of the village now stands at around 1200.

These new houses have attracted affluent commuters and retired people. In fact we can identify three social groups. Post-war council housing effectively consigned the working class to the very outskirts of the village (symbolically the allotments are down "that" end). The new estate brought in people in the middle income range, while the new housing in the village heart went, as one incomer put it, to those in the "upper managerial" strata. The development processes set in train important social changes. The result has not been simply that of a division between "incomers" and "locals", for there are also divisions between the incoming groups. These divisions have become clearer as these groups have attempted to forge a new identity for the village. This has taken place most notably in the village institutions.

There are between 20 and 30 formal groups and societies operating in the village. These include an art club, a garden society, a music group, a playgroup, the ramblers, the Conservative Association, the Sunshine Club (for the over 60s), a toddlers group, the Wednesday Group (various activities), the Wingrave Players (drama), the Wingrave Singers, "Music from the Vale" (recitals), and a Twinning Association. Many of these date from the establishment of a Community Association in 1973 which sought to "promote and further community activities within the village" (Wingrave Directory). However, not all "sides" of the village participate equally in these activities. As one long-term (20 years) resident put it: "Wingrave has a very active community spirit but sadly it's nearly all new Wingravians rather than the old . . . . Basically the Community Association is run by incomers". Another resident (in the village 12 years) commented: "The people who run the village have arrived in the past 20 years. The type of person who moves in is perhaps more inclined to do things". This, he went on, had resulted in "a 'them' and 'us' situation" between the two ends of the village. A recent entrant to the older end of the village elaborated on this: "There is a split in the village – I'm tempted to say a class split – between those whose expectations of life are more (newcomers) and those whose life expectations are less (locals) . . . . The

113

people down 'that end' don't join in as much as people 'this end'. They do their own thing which is much more family centred". This situation prevails not only in the social organizations but also on the Parish Council. As one Council member explained: "I've been on the Parish Council 13 years. People on the Council have changed. When I first came on I was one of the first newcomers, the remainder had been on there 20 to 25 years, local people. Gradually over the years the old guard disappeared and the council changed to almost all incomers" However, "there is a feeling between the two sides of the village – we've tried to get a Parish Councillor from 'that end' [council house/new estate] but have not succeeded". So the Parish Council is made up of newcomers, i.e. from those who have moved into the old part of the village.

As well as the social marginalization of certain incoming residents what is also noticeable is the increasingly marginal position of farmers and farming. There are no farmers on the Parish Council and very little farming involvement in the Community Association. Those farmers remaining in the village seem quite alienated from village society. As one put it: "I used to be able to drive my sheep through Wingrave when I wanted; now I'm just a public nuisance . . . I have very few close friends in the village, I have good friends on neighbouring farms. People in Wingrave are predominantly 'yuppies'. I used to be on the Community Association Committee but I got sick of the people who moved in. The people who moved in don't really understand farming life; I can't get on with them"

According to the District Councillor "Wingrave is becoming less of a village, more of a dormitory settlement". However, it is quite clear that some of those who have recently moved in are making a concerted effort to maintain or develop some form of village "community spirit". One incomer explained: "We moved here from an executive dormitory and threw ourselves into local life in various ways because *that's what we wanted from the place*. We needed to build up severed relationships and we were very well received here; this is a very friendly village" (emphasis added).

There is also some evidence that people are moving to Wingrave in order that their children can have a "rural" education. As the village has grown so has the school, from 70 pupils in 1961 to almost 200 in 1991. Increasingly the type of child attending the school has changed. As one teacher put it: "Its mostly commuters' children. There are no farming families sending their children here. On County Show day we always used to have children off but not now. We get children from Aylesbury, as parents choose this school over the town school". One reason why the school is popular, the teacher believed, is because "our children are pleasant, easily managed, friendly, open; we've got only one child on free school meals, on income support". The school then, is in the main, a "middle-

114

class space" where children are seen to get a "good" education. However, another reason why the school is popular, at least with some parents, is the absence of ethnic minority children. According to the teacher, some parents living in Watermead were sending their children to this school, rather than to Aylesbury schools, for precisely this reason. As he put it: "It might be racist but if your children are going to a class where to half the children English isn't their first language and a lot of attention is paid to their cultures you might wonder yourself whether your child might be better off somewhere else". Thus, Wingrave is attractive to many middle-class parents as it offers the "right" (white?) environment in which to bring up their children.

As these patterns of social change impose themselves on the village it becomes reconstituted as a social space. Although this takes place in complicated ways it arises in part as a consequence of Wingrave's lost rôle as an economic space. Farming has almost completely deserted the village and resides in the surrounding open land and farmsteads. This "desertion" has opened up room for the incomers to stamp their identity on village society and in this sense the village is still in a process of "becoming". What the village will "become" is unclear but it is likely to bear more than a passing resemblance to some form of middle-class "rural idyll", with a considerable percentage of the population, particularly those on low incomes, marginalized both socially and spatially from the village "heart". Nevertheless, this "idyll" will only come into being as a result of some kind of negotiation between the various social strata. The recent in-migrants cannot simply ignore other groups, and they must somehow find ways of accommodating these "others" within village society. Indeed, the ascendant group displays a conspicuous desire to do this. Local identity runs parallel to class identity.

The rate of change in Wingrave has been more rapid than at any time in its history (see Fig. 4.3) and there was a feeling in the village that if the community was to come to terms with these changes then development pressures should be eased. As a member of the Parish Council argued: "One of the most critical things is the rate of change. You actually have to pace out development because the physical infrastructure needs to cope but so does the community . . . . The village has been able to cope until now. There is now the need for a breathing space; we need a slow down". Recently two parts of the old village have been given Conservation Area status. However, this designation does not halt development completely and within the village there was some scepticism as to whether it would have any effect at all. The District Councillor commented: "The Conservation Area designation has made little difference. They [the planning officers] draw up these wonderful plans but don't seem to take much notice of them". However, it seems likely that where there is development within the old village it will be of the type considered in the case

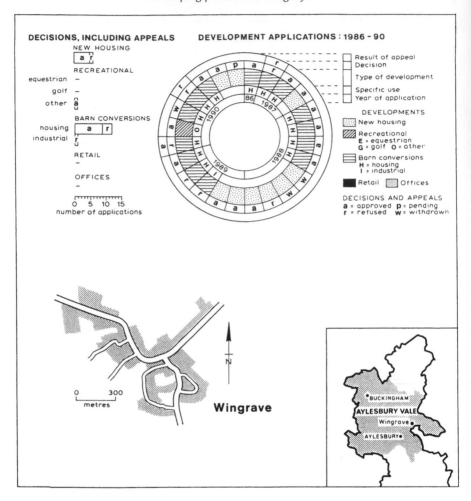

Figure 4.3 Development pressure, Wingrave, 1987–91.

study in the previous chapter: sympathetic construction in traditional materials, and therefore expensive. Conservation Area status may well distinguish still further the two "ends" of the village, both aesthetically and socially.

In Wingrave we found that the fracturing of the population into discrete segments was particularly evident. The population growth experienced in the village since the 1960s, as well as the more recent changes in the centre of the village meant that, in contrast to Swanbourne, different groups in the village were competing for village resources often expressed by differing representations of the village, and – as with Weston Turville – sometimes focused upon the changing rôle and use of central village facilities (village greens, public recreational space and other village amenities and institutions). Such concerns were particularly prevalent around land development proposals, notably the semi-privatization of former public spaces. For instance, Wingrave contains a large village green and a recreational site, both of which had originally been bestowed to the village by the Rothschild estate in the late 19th century. These are central features of the village. Access to these sites, however, is becoming reorganized by dominant groups. In interviews with a range of residents we found that the use of central village recreational facilities had became a major issue which respondents used to characterize the nature of social change.

For example, an application for a new recreation ground (a two-stage development of a recreational park) was made by the Chairman of the Parish Council in the late 1980s. The Parish Council had taken the opportunity to lease (for 21 years) the 3.5 ha of land from a local landowner in a field adjoining the back gardens of the council tenants. With strong support from the recreational committee, it was argued in the application that because Wingrave had doubled in size over the past 20 years the existing central village facilities were becoming overused (this related particularly to football [with six village teams] and cricket) and there was a need for more formal recreational provisions. Two new football pitches with changing facilities were to be established by September 1991, to be followed, under a second application, by a pavilion development and 50 car parking spaces. Permission for the first phase was granted in March 1990, conditional upon appropriate tree-planting and landscaping criteria designed to "shield" the council estate.

There was little formal inspection of this application from village residents, with only four letters of objection received by the planning department. Most of the issues concerned siting details and the inadequacies of the physical site. Only one council tenant formally objected, arguing: "What a waste of money to have another sports area. Some of our youngsters would like homes". The planning permission was officially justified in the following way: "In the light of agricultural set-aside policies it is considered that there can be no objection to the proposal on the grounds of loss of agricultural land. In landscape terms a playing field can be acceptable within an area and car parks are to an acceptable scale, sensitively sited and detailed."

These issues were particularly prevalent when we interviewed a random sample of council tenants. Sited on the periphery of the village, but adjacent to the new recreational site, this group of residents were highly critical both of the long-term changes that had taken place in "their" village and the more recent reorganization of social institutions. We include here four cases from our interviews. All were residents of the council estate on the perimeter of the village and none of them had purchased their homes under the "right-to-buy" scheme. The estates were built in the 1950s, comprising of two adjacent terraces leading to a cul-de-sac overlooking open fields.

The self-employed carpenter　This council resident was originally from a village in the Cotswolds, while his wife and live-in father-in-law were born and bred in Wingrave, their family extending back over many generations. Regarding the main land development changes he cited the "new estate next door" which was highly visible from his workshop/ garage and the new barn conversions in the centre of the village.

He argued that the level of new development over the past 15 years had "taken the character out of the village, it is now 'them and us' . . . there always seem to be splits". The new estate had been started 10 years ago and finished only two years ago. He believed that "new communities" should be built in rural areas rather than spoiling old ones. Wingrave used to be a close community – but not any longer: "People come along with different ideas about how the village should be run. Also they have taken over the Parish Council, largely because of local apathy".

He referred to the changing use of the old recreation ground in the centre of the village. It used to be informally run, with people using it when they pleased on a "give and take" basis. "The newcomers come and want to reorganize it, and then even develop a new recreational area as well." The latter is just, he argued, a "dogs playground; they have leased the land from the charity for a dog park". The old recreation ground was public space, originally owned by Rothschild who bequeathed it to the village in the 19th century. It was never meant to "be a source of conflict". This is all due to the "growth of the village".

He argued that the small council estate tends to have the more established village residents with many of them going back generations. People moved from the centre of the village to the estate in the 1950s. Now the farmers are moving out too. Nevertheless, there were a lot of young people who wanted to stay in the village but could not afford it. He estimated that 70–80% of the young want to stay after they have married. He complained of there being no access for young people: "They have to go to Aylesbury or Leighton Buzzard to find homes". Also he argued that generally 75% of newcomers' children were also attached and wanted to stay, but they could more easily afford to.

As far as participation in village activities were concerned he was a member of the village Bowls Club but that was all, although he did do some voluntary work at the community centre. He confirmed that farmers were a less influential group now and said there was "no rôle" for them in the village any more.

Regarding barn conversions, he thought that those at "Essex Yard" (see Ch. 3) were well done, but "some further down the road" were really a new housing development – i.e. hardly any remnants of the barn(s) remaining. He did not feel as opposed to the barn "conversion" as he did to the estate-like development next door. In fact, as a carpenter he said, "the changes have been good for me". Over the past ten years, except for a brief period, he has found all his work in the village, largely from amongst the more established residents pursuing renovation etc. He is thus an active agent in the reconstitution of the village, yet despairs at some of the consequences.

He did not envisage moving, mainly because his wife was very attached to the village. His social and economic life was very much village-based. He argued that the new estate (next door) was very socially self-contained. There were lots of social networks in the council housing area, many of them kin-related. Many of the older residents had, he said, formerly lived in "two-up-two-down" accommodation in the village heart. "Now the village has lost its identity. Something has gone. Newcomers come in and want the village in their own shape".

Young single mother This respondent had recently moved to Wingrave from the nearby village of Wing, which she believed had a much more lively social life; it was where her family (and boyfriend) were located. She had been allocated a house in Wingrave after living in overcrowded conditions in Wing.

She regarded Wingrave as a rather soulless place. There were few shops and a car was needed for getting around; "I can't just pop out to the shop. I don't have access to the car every day". The playgroup in Wingrave was regarded as hostile. It was largely disorganized with a few "other women setting the rules". This resulted in her taking her child back to the Wing playgroup when she had the transport.

She preferred to live in a village – she saw "village life" as better for children – but did not see herself as part of Wingrave. As far as she was concerned there were not sufficient recreational facilities. More houses were needed, she felt, as she had to wait two years to get a house.

This woman felt out of touch and excluded from the main social–recreation facilities, but was generally supportive of protecting the village against new developments.

Retired woman She had lived in the same house in Wingrave for 22 years. This woman had a very dense local kin network. One son (in his 20s) lived with her, two sons lived next door and two daughters lived in Wing. She had 27 grandchildren.

She was very aggressive when it was suggested to her that there seemed to be a lot going on in Wingrave: "There are no shops, only one

119

in the village. The butcher was done away with and they closed the Bell Pub. There's nothing for the younger people. They have to go to Aylesbury. There are quite a few young people here, but the Youth Club closed down."

Housing was seen as a difficult issue – and was given as a reason for her son residing with her. She advocated the need for more shared ownership schemes; there was "no chance of him [her son] getting a house anywhere". There was not even any access to rented accommodation. This meant she said, that "the youngsters will move away somewhere" Furthermore, "there's no work around here."

In terms of the village itself she argued: "The top of the village, they've got the houses, we're council down here. We've been here 20 years, they come here and they're on all the committees. The community centre was open (the old school) for ten years. They started renting it out – it got more expensive, and now you can't afford to hire it. There's recreation up there: tennis, but you have to "join the club" and pay – we used to go and play when we wanted. There's nothing for the little ones" There were more children (and therefore "need") "down here" in the council houses, but all the recreational facilities were in the top end of the village.

What did she think of the planned new recreation centre (at the back of her house)? She was not involved in any of the planning processes. Despite having 27 grandchildren in the area, she felt aggrieved at this. It was literally in her back yard. She said that she had seen a "man in a white car stop and inspect the site, that's all I know." A lot of people had put up a petition "round here" not to have a recreation ground. "Older people didn't want it because of the kids, but they have to go somewhere. It's going to be a football pitch, cricket pitch, then they will have a pavilion – then they'll begin to charge. The cricket club, football club pushed to have this second pitch."

She travelled to Wing for prize bingo every first Monday of the month – "the place gets packed and the proceeds go to all the old in Wing. There's nothing like that here. They're just not interested. There's a school harvest festival in Wing and the boxes of groceries go to all the old people." Buses were infrequent to Aylesbury. The young "hang around" at the top of the village. She did not feel that there was sufficient entertainment in the village for her either.

This respondent felt a strong need for more "simply organized" forms of activity in the village. She expressed frustration both for herself and the family at the way the village was run. Her feelings of exclusion were emphasized by the development of the new recreation ground right outside her back door.

Middle-aged couple with two adolescent children at home; father unemployed due to ill health   The wife was from the village originally; they had lived in the same house for 17 years. The family had noticed considerable change in the village. There used to be two doctors' surgeries, now you have to go to Wing. The daughter (18) had arthritis. "There are no relevant clubs or activities in the village". In one sense there is a lot going on (drama clubs, etc.) and a community association, "but not really for us". There used to be four pubs, but now only one, and two shops, a butcher and a bakery. All these have gone over the past 20 years. Now they have to go to Aylesbury.

"The village is divided into two from the Old Bell pub". Beyond the pub was seen as a different village. There was redevelopment and new housing in this other part. That is where the development was seen to be. This was seen as a negative feature of the contemporary village. It was felt that "village life should be one big community". However, "new people take over everything: the community centre, recreation ground" Again, there seemed to be an overriding feeling of exclusion.

The husband was very much a social person. He had started a one-man disco club in the community centre "laying out a lot of money. It was specifically designed for the young but it fell through. Nobody came to it". It cost then (about ten years ago) £25 per night to hire the room – so "I had to charge an entrance fee". Now "we hired it for our daughter's 18th birthday party (recently) and it cost us £75". The youth club closed because of lack of funds.

Planning permission had been granted for the new recreation ground. There were lots of different football teams and few pitches. Its development reflects the rising activity for football, cricket and tennis. The footballers were seen to be the main actors in forcing the Recreational Committee to find an additional site. "'Who is on the Recreational Committee?'", these respondents asked. They seemed quite perplexed at the recreational ground development. They thought it was likely that, although they lived adjacent to the site, they would have limited or no access to it once development had taken place. While this may have always been true with the original farm land, given that it was now to be seen as a new recreational site, this caused concern.

These interviews indicate some of the key consequences of land developments which might be described as the twin processes of middle-class mobilization and the social fracturing of village life. These were particularly evident in Wingrave. The council tenants were physically and socially removed from the formal organizational activities in the village centre, yet their historical knowledge of the village and their kin ties attached them to the place. The privatization and social reorganization of public village space, and the development of the new housing estate, acted to emphasise their distance from the "core" of village life.

In Wingrave, probably more so than in our other two villages, we see the potential internal conflicts which arise not only from the attainment and maintenance by some of their positional goods (i.e. housing and its immediate environment) but also the fine-grained conflicts and fissures which develop around the social organization of space, facilities and amenities as land development processes unfold. The expansion of recreational facilities in Wingrave, while highly relevant to the village as a whole, was being pursued by certain social groups and constructed around their needs and priorities. Sporting activities, together with a whole host of indoor clubs and societies, provided the very active appearance of the village, but certain groups felt excluded from these activities because their organizational needs were not being met and often the costs were prohibitive. The re-use of the community centre, the village green recreation site, and the conversion of former agricultural land for a new recreation park effectively redistributed the social rights within the village to certain dominant groups. In Weston Turville we saw a clearer version of this concerning the proposed development of glebe land. A key issue here concerned *which* social actors – established middle-class residents, landowners or farmers, middle-class incomers, council tenants, etc. – had the relative power to rearrange the public and private use and occupier rights. In villages such as Wingrave and Weston Turville, where populations have swollen, these problems are constantly confronted not only concerning economic resources, but also over the ability to act, represent and organize the maintenance and development of the village. In Wingrave our interviews with different residential groups (council tenants, owner- occupier housing estate residents, village centre residents) point to the differential power of actions and representations around different development issues.

Conclusion

In the previous two chapters we have seen how actors in the land development process are working with different conceptions of what "the rural" is and should be. Watermead, for instance showed how a developer can, seemingly arbitrarily, take elements from both the urban and rural domains and subsume them within the setting of a new village. The juxtaposition of these elements does not seem unduly to concern those people who consume this space. They see this as a unique environment which facilitates a particular kind of lifestyle. This was one response to the desire to live in a village within the context of restricted access to the "real thing". The developer, having negotiated building rights on a parcel of (almost rural) land sought to attract people to a new village location. The scale of this development (800 houses) would, the planning authority hoped, take some of the pressure off the other established villages. Presumably, therefore, Watermead was designed to appeal to those who sought a village location, and there is some evidence that this is the case.

The different local conditions suggested by these village studies indicate that there is not one "culture" associated with the middle class in the rural areas of Buckinghamshire, although we would agree that these "cultures" in general are becoming "hegemonic". Opportunities have arisen, due to the decline in agriculture and the social groups associated with the industry, which have allowed the "new" arrivals (stretching back over the past 20 or 30 years) to "colonize" these spaces. However, *the cases studies indicate that there is no unified culture associated with* these arrivals. Watermead, for instance, represents a younger, more urbanized, aspiration for rural living (the developer believes they are seeking a media image "come to life"). Wingrave, on the other hand, shows how, incrementally, the different types of development can result in the arrival of different social groups, seemingly with different expectations and aspirations. Here any semblance of a unified culture within the village is only coming about in a partial and uneven fashion, as the most affluent attempt to forge a new community consciousness. What both Wingrave and Watermead have in common, however, is the desire for a particular kind of living space, a space which extends beyond the house itself out into the surrounding environment. This environment is physical (aesthetically pleasing buildings, open space, countryside) and social (village life, being part of a community, etc.). We would contend that those who seek such environments believe them to be easier to attain in the rural arena. This of course depends in part on their experience of the urban arena.

The villages, and the development processes which shape them, offer contrasting ruralities. They are seen to represent authentic rural locations. In Swanbourne the traditional social and material shape of the village remains intact, while in Weston Turville it has almost completely disappeared. In Wingrave there is a struggle for authenticity. This struggle is being spearheaded by the most recent arrivals in the village, those who have moved into its most authentic part.

While we find a range of actors in these settlements utilizing various conceptions of rurality (often very locally specific, based around ideas of the neighbourhood) we believe that the notion of class allows us to weave these conceptions and concerns into an account which shows how they merge to become key forces determining the course of local development processes. The outcomes of these development processes can be considered within the terms of the broad categorization that "class" allows. Within this we can identify many elements such as length of residence, degree of attachment to both community and place, property rights, provision of facilities, desire to leave the "urban", etc. If we see these as providing the characteristics of a certain sort of social space then we can begin to understand how this might be a *class* space. The villagers of Wingrave, for instance, whether they live in the council houses, on the new estate or in barn conversions, are all acutely aware of the fragility of "their" community. And this is because the community is increasingly taking on particular characteristics as the middle class forms itself in that place. It thus continually excludes some who find it hard to identify with the practices and rituals of the community institutions (i.e. those resident on the council estate) while others (i.e. middle-class in-migrants) recognize themselves immediately in these ways. Yet all live within one small settlement and so, to some extent, are aware of the existence of "others". Middle-class formation is incomplete in Wingrave, but any observer of the village's history would clearly be able to see the process as well advanced. In Weston Turville, on the other hand, the rate of change has been so fast and the size of the numbers moving in so large, that the identity of the place has almost been lost. Furthermore, it is hard for the incomers to reconstitute the village in any meaningful way for they are surrounded by many people who have also just moved in. So all are in flux and the village as a whole has an air of social impermanence. Allied to the threat from Aylesbury, this leaves Weston Turville unsure of its future, although many residents appear willing to fight to retain the "village". If the rate of change slows in forthcoming years then a more vibrant and cohesive community may emerge. However, at present, although there is a strong middle-class presence in the village it has little coherence.

Thus, when we characterize Buckinghamshire, or Aylesbury Vale, or Wingrave as dominated by particular class conceptions of rural space we are not seeking to impose some simplistic, spurious, homogeneity on these areas. Rather, we are attempting to identify the key processes of change, and the key participants in these processes, which *are* determining their material shape and social complexion. The notion of class provides a useful way of integrating actions within the development process, social change in discrete areas, and conceptions of rurality. The development process is crucial to these wider processes of change. As we have seen, the housing development process delivers particular types of housing and thus particular social formations. Once these come into being they ensure that only certain sorts of housing get built (that is if they get built at all). Thus, the process becomes cumulative. The planning system structures these processes and provides resources for those actively seeking to mould these spaces in particular ways. Thus far, this class action has been focused on rural settlements where, by and large, middle-class formation and planning policy dominate the development process. However, as we show in the rest of this book, these forces are now reaching out into the wider countryside.

CHAPTER 5

Agriculture and beyond

Introduction

The village studies indicate that in some settlements, such as Swanbourne, agriculture still plays a central part in community life, while in many others, such as Wingrave and Weston Turville, it is utterly marginal as farmers have "retreated" to the open countryside. This remains their "domain" and agriculture is still the major land use in rural areas. However, even in the open countryside, agriculture no longer "reigns supreme". As we outlined in Chapter 1, the planning system has, since 1947, followed a policy of "urban containment" and has, therefore, been in effect a system of *settlement* planning. Agricultural land has traditionally lain outside any kind of planning control. However, a profound shift in this post-war arrangement can be discerned during the 1980s. There are two main reasons for this: first, attempts by successive Conservative governments to recast the rôle of rural planning; secondly, the crisis in agriculture. We shall here briefly review these two broad changes before turning later in the chapter to identify their effects on agriculture in Buckinghamshire.

In *Constructing the countryside* (Marsden et al. 1993) we outlined the changing nature of rural planning, showing how, during the early to mid-1980s, the government took steps, in line with its general policy of "liberalization", to reduce the scope of planning. However, these moves in the rural sphere were much less interventionist than was evident in the urban realm and ultimately, under pressure from local authorities and rural interest groups, the opposite result was accomplished; the planning system was stronger by the end of the decade. For instance, in the early 1980s it was proposed that structure plans should be abolished. However, by the early 1990s a major rethink had taken place on this proposal and, although there were changes in the format and scope of such plans, they were now encouraged to play a much greater part in rural areas than previously. For example, in DoE Planning Guidance Note 12 published in 1992, eight themes were outlined for inclusion in structure plans, four of which have direct relevance for rural areas: green belts and conservation in town and country; the rural economy; mineral working and protection of rural resources; and tourism, leisure and recreation.

Furthermore, under the 1991 Planning and Compensation Act district-wide local plans were made mandatory for the first time. The effect of introducing development plans covering the whole of each district was to bring all rural areas under the local development plan-making process. Indeed, the spread of district-wide plans, a commitment to county structure plans, a revival of regional planning, and the continued reliance on localized special area plans (to cover, for example, expanding towns), as well as the requirement for county councils to produce local plans on minerals and waste disposal, have led to a proliferation of plan-making activities (Lowe et al. 1993).

In the light of the analysis presented in Chapter 1, it would be reasonable to assume that the increased scope of development plans implies an enhanced rôle for the middle class in determining the scale of development in the countryside. But before making such an assumption we must consider the future status of agriculture, which for so long has lain outside the planning system. Moreover, the effectiveness of middle-class groups is likely to vary over time, from locale to locale and from sector to sector. In subsequent chapters we examine the rôle of the these groups in different sectors of the development process, but in this chapter we concentrate on agriculture.

Agriculture in the rural economy

During the 1980s agriculture entered its most serious crisis of the post-war period. It is commonplace to argue that the sector has been a victim of its own success. In the post-war period the aim of agricultural policy was to modernize the industry in order to boost output and productivity. In this it has been successful: labour has been shed; capital investment and the use of technology have increased enormously; and food shortages have given way to food surpluses. It is perhaps not surprising that a political storm blew up around the industry in the 1980s as enormous amounts of taxpayers' money continued to be diverted to the sector at a time when markets could not be found for many of agriculture's major food stuffs. Furthermore, the environmental consequences of modern agricultural practices, such as pollution and landscape degradation have caused growing disquiet amongst the public at large. If we add to this the part that agriculture plays in an increasingly complex food system (Goodman et al. 1987, Marsden & Little 1990), in which farming represents (in terms of the proportion of the value it adds to the final product) a shrinking sector and where farmers find their autonomy significantly reduced, we can begin to see the full dimensions of the crisis facing the industry. Agriculture is now likely to go through a period of profound readjustment. However, the depth and severity of this, as well as its duration, depend, to a large extent, on the future course of agricultural policy.

According to Harvey (1993) agricultural policy reform (within the Common Agricultural Policy [CAP]) has been poised between: first, the direction of increased supply control and isolation from world markets, as illustrated by milk quotas and other similar measures; and secondly, the exposure of the industry to world prices and "de-coupled" support, as demonstrated by the MacSharry proposals and the GATT negotiations, whereby subsidies are no longer linked to output. At present it is unclear which of these is the most favoured route for policy makers, but budgetary constraints are likely to continue to be imposed and international pressure to reduce subsidies to farmers will be maintained in multilateral trade negotiations, forcing cuts in price support. What is also not clear is the effect this will have on British agriculture. For instance, commentators disagree on such basic issues as the likely levels of resource use in the industry. North (1990), for example, believes that the amount of land currently being used in agriculture will fall as prices fall. He calculates that, by the year 2015, five million ha will go out of agricultural production in the UK. Harvey (1993), on the other hand, claims that very little land will leave agriculture. He says: "suppose that the reduction in output necessary for the sustainability of the CAP or the countryside is of the order of 20%. For the sake of simplicity, it can be assumed that this reduction could be achieved through the release of either labour or capital from the industry, rather than land. In the context of historical changes in the agricultural industry, changes of 20% in the labour and capital employed in the industry are commonplace over relatively short time horizons" (1993: 14). Harvey sees little likelihood of land becoming surplus to agriculture, especially as under present policy arrangements compensation for price reductions and other forms of support is tied to farmland areas. Ultimately, land will only be released from agriculture if suitable non-agricultural land activities provide sufficient incentive for landowners to shift uses or agricultural returns are driven so low as to make agriculture uneconomic. The latter scenario is unlikely, given the level of political commitment to the farming sector. To understand the scope for the former will depend to some extent on the limits placed on land transfers by the planning system.

One response to the agriculture crisis favoured by the Ministry of Agriculture has been to promote diversification of the rural economy (Shucksmith & Winter 1990). Diversification was seen as a means of "weaning" farmers off price support as they came to supplement their income from non-agricultural sources. Thus, public subsidies could be reduced and budgetary pressures eased. In line with these, and with increased environmental concerns, the Ministry began to shift its focus away from agriculture *per se*. In the Agriculture Act of 1986 the remit of MAFF was widened to include the following considerations: (a) the promotion and maintenance of a stable and efficient agricultural industry, (b) the economic and social interests of rural areas, (c) the conservation and enhancement of the natural beauty and amenity of the countryside, and (d) the promotion and enjoyment of the countryside by the public (Hodge 1990). In line with increased support for diversification, MAFF began to withdraw its veto over non-agricultural transfers of farmland. Under the DOE Circular *Development involving agricultural land* (1987) it was suggested that development proposals should only be referred to MAFF on the higher grades of agricultural land. Thus, it was expected that the number of MAFF and Welsh Office consultations would fall from 1400 in 1988 to a maximum of 200 (Marsden et al. 1993: Ch.5). However, the withdrawal of MAFF from the planning process would not necessarily ensure that farmers would easily be able to diversify, for they would now have to negotiate their way through the planning system.

The rôle of planning in promoting or resisting diversification of the rural economy has fluctuated with the efforts initially to free up and then subsequently to strengthen the planning system, as referred to above. Part of the rationale behind the Conservative Government's early efforts to weaken the rôle of planning was the hope that rural enterprise could be promoted, particularly amongst farmers. However, under pressure from amenity and conservation interests the concern to protect the countryside eventually became paramount. In the Planning Policy Guidance Note [PPG] *The countryside and the rural economy*, published in 1992, it was not agricultural land but the countryside that was now to be protected "for its own sake". Any non-agricultural developments would have to be assessed in the light of aesthetic and environmental criteria. Furthermore, pressure was growing to subject agricultural activities to planning decisions. Hence, the PPG states that: "While the government has no present plans to extend planning controls to all farming activities, it is ready to introduce new closely targeted controls where this is necessary to deal with specific problems" (para 22). Thus, planning would seem to be on the verge of achieving another extension of its remit in rural areas.

In Chapters 6 and 9 we examine the impact of this shift in planning policy in the context of specific development processes resulting from attempts to diversify. However, in the rest of this chapter we examine how farmers in Buckinghamshire are dealing with the problems thrown up by the crisis in agriculture, focusing specifically on the development and management of land.

Agriculture in Buckinghamshire

We begin by outlining the broad patterns of development in the local context of Buckinghamshire. These are identified over a twenty year period for, as we have already mentioned, farming has recently entered a period of crisis. It is first necessary to give some indication of how policy and market pressures have become manifest in the locality. In what follows we outline the broad patterns of change in Buckinghamshire (using MAFF Agricultural Statistics). We then turn to examine how land is being utilized by farmers in the restructuring of their agricultural enterprises in an attempt to understand whether land is likely to be lost from the industry in the future. We use survey material to show the extent of land transfer and case studies to understand the decision-making processes.

Farming in Buckinghamshire has traditionally been mixed – mainly dairying, sheep and arable. The quality of the land partially conditions this farming pattern, the majority being in Grades 3 (67%) and 4 (23.9%) with smaller proportions in Grades 2 (8.3%), 1 (0.7%) and 5 (0.1%). The average quality of the land has however not prevented the pattern of farming from being profoundly affected by the various permutations of policy in the post-war period, for farmers have generally been able to adapt the land to whatever requirements have been paramount. This can be seen in a brief examination of agricultural change in the period since the UK entered the European Community.

Between 1973 and 1989 trends towards specialization have been evident in Buckinghamshire, most notably a shift to arable farming and a reduction in permanent grassland. The amount of permanent grassland declined by 20.4% and was replaced not by barley (–52%), but by wheat (+57%) and oil-seed rape (+312%). The increases in these two crops reflect shifts in agricultural support under the CAP. With a deficiency of vegetable oils within the EC, attractive prices were offered for oil-seed production. The rise in world commodity prices during the 1970s also encouraged production and the crop expanded mainly at the expense of barley. Most noticeable in the livestock sector between these two dates is a decline in dairy cow numbers (–24.5%), particularly after 1985, the year milk quotas were introduced. Pigs also declined steadily throughout the whole period (–47.3%), reflecting the concentration of the pig industry in the East of England. Sheep, on the other hand, increased by 53.3%, despite a decline in the mid-1970s. This increase is particularly marked in the late 1980s and can be attributed to the shift out of milk and the relative profitability of the EC's sheep-meat regime for lowland farmers.

Policy has not only affected the pattern of cropping and husbandry on Buckinghamshire farms but also the size of farms and the farm labour force. The changing size of holdings between 1973 and 1989 is most evident in the middle-range categories. The smallest holdings (below 20 ha) have remained relatively stable, as have those in the largest categories (> 330 ha). In the middle range categories the holdings in the 100–299.9 ha category gained at the expense of the 20–99.9 ha category. The former declined by 27% while the latter increased by 22.5%. By 1989 the larger of these two size categories accounted for 18.4% of all holdings while the smaller accounted for 37.6%. This compares to 44.3% and 12.3% respectively in 1973. The effects of policy and technological change have also combined to a marked degree in shaping the labour demands on Buckinghamshire farms. The sharpest decline here has been among full-time regular workers (–46%), followed by regular part-time workers (–21.3%). Seasonal and casual work has also declined (–21.3%) but the numbers involved were not large. The only category which has increased is the farmers, spouses and managers category (+3.2%). This can also be linked to the shift in tenure during the period, whereby the ownership of holdings has increased (+16.7%) at the expense of those rented (–50%) and those under joint tenure (–35.1%). In 1973 owner-occupied holdings made up 44% of the total but by 1989 this figure had increased to 63.7%. The number of rented holdings declined from 29.8% of the total in 1973 to 17.2% in 1989. The shift to owner-occupation has thus been marked during this period.

The picture here revealed by MAFF statistics is not too dissimilar from what we would expect from national trends. There has been a move to slightly larger and fewer holdings, with a sharply reduced labour force. The intensity of production has arguably been increased by the shift to wheat and oil-seed rape, while the decline in dairying has led to a large increase in sheep numbers. Farming *sensu-stricto* has therefore changed its shape during the 1970s and 1980s largely along productivist lines, promoted by the support systems of agricultural policy. This is evident in the very gradual decline between 1973 and 1989, in the farmed area of only 7,701 ha (or 5.5%). The amount of land in agricultural use has remained remarkably stable, given the intense economic and social changes taking place in the county.

Aggregate figures such as these tend to give an impression of smooth and unproblematic change. However, as we have already indicated, from the early 1980s onwards the nature of EC and national level policy was significantly modified. Lower support prices, more uncertain market conditions and a return of positive interest rates, with steady or declining land values, generated considerable uncertainty for farmers. Falls in capital investment and the need to create sufficient working capital so as to replenish the farm business, led to the perceived need to diversify incomes, assets and types of production. The onset of these more uncertain conditions is expressed in changes in the rôle of land, and in farm decision-making. We can begin to assess the reaction of farmers in Buckinghamshire to these new pressures from the longitudinal survey of 227 Buckinghamshire farmers undertaken as part of a Europe-wide survey funded by the European Commission (see Appendix 1). In the sample, following consistently the MAFF census statistics, 62.8% of all land was owner-occupied, with 29.1% under conditions of full tenancies. Of the rented land, however, only 49.1% was under full agricultural tenancies.

130

The majority of tenants (50.9%) were engaged in a variety of short or medium term arrangements i.e. short lets, landlord–tenant partnerships and share-farming agreements. While these still represent a minority of those on tenanted land, they cover more than half of the tenants in the sample. The survey was undertaken in 1991 and we asked the farmers whether, since their 1987 production year, there had been any additions or disposals of land to or from the farm (including land rented in or out). The results are given in Table 5.1.

Table 5.1 Farm survey: land additions or disposals (1987–91).

	%	Numbers
Additions only	8.0	18
Additions and disposals	2.2	5
Disposals only	15.0	34
No change	73.8	167
Don't know	1.0	1
Total	100	224

The majority of farmers had not acquired or disposed of land during this relatively short time period, which is much as one would expect. Of those who had engaged in some form of land trading, the majority disposed of land. The results indicate that over a quarter of the sample had engaged in some form of land sale or purchase. Evidence was also available on the methods used for the acquisition of land and the motivations for doing so. The results are given in Table 5.2, where two categories – those of outright purchase and renting – predominate. This is borne out by the main reasons for acquisition, given in Table 5.3.

Table 5.2 Farm survey: method of acquisition (for those acquiring land, 1987–91).

	%	Numbers
Inheritance	8.3	2
Lifetime transfer of tenancy	3.1	1
Purchase	41.6	10
Use (no payment or ownership)	2.1	1
Acquisition of operation through management	4.1	1
Rented in	41.6	10
Total	100.8	25

Clearly farmers and landowners were taking in land as it became available, either for purchase or for rent. Those disposing of land also provided information on method and reasoning (Tables 5.4 and 5.5) showing that the disposals mainly involved land sale, although there was some amount of renting out. The need for cash was invoked as the overriding reason for disposing of land, and this indicates that financial pressure was forcing the use of land as an asset capable of relieving such pressure. The bulk of this land went to other landowners for agricultural purposes.

Table 5.3   Farm survey: method of disposal (1987–91).

	%	Numbers
Lifetime transfer of ownership	5.4	2
Lifetime transfer of tenancy	2.7	1
Sale	78.3	29
Rented out	13.5	5
Total	99.9	37

Table 5.4   Farm survey: reason for disposal (1987–91).

	%	Numbers
Needed cash	30.0	12
Transfer to family	2.5	1
Could not refuse offer	15.0	6
End of tenure	27.5	11
Compulsory purchase	7.5	3
Other reasons	17.5	7
Total	100	40

Table 5.5   Farm survey: planning to take land out of agricultural production (average ha per respondent in parentheses).

	%	Numbers
Set aside (139)	39.2	11
Idle land, not set aside (1.25)	7.1	2
Woodland/forestry (15)	32.1	9
Erection of buildings (1.5)	7.1	2
Land for other non-agricultural purpose, e.g. sports/ recreation (30)	39.2	11

Table 5.6   Farm survey: reason for acquisition (1987–91).

	%	Numbers
Lifetime inherited	8.3	2
Land became available	66.6	16
Availability of labour	8.3	2
Need for more land	16.6	4
Total	99.8	24

The survey asked whether, since the 1986–7 production year, there had been an increase in the utilization of land for non-agricultural purposes (including land cultivated in the 1986–7 production year but now left idle other than as part of a crop rotation). Those answering "yes" numbered 28 (12.7%), while those answering "no" numbered 194 (87.3%). The former group were asked to detail what this "non agricultural" land was used for. The answers are given in Table 5.6 and show that the bulk of this "non-agricultural" land went into set-aside, usually on the larger farms. The next highest proportion, however, went into activities outside agriculture such as sport and recreation, the most extensive of the non-agricultural land uses. Farmers were also asked whether they had plans to take land out of agriculture during the forthcoming five years. In answer to this question 31 said "yes" (13.6%), 169 said "no" (74.4%), while 27 (12%) did not know. Of those saying "yes" the uses to which the land might be put are given in Table 5.7 and show that land for set-aside is the most popular "non-agricultural" use, with sports and recreation and new buildings following on.

Table 5.7 Farm survey: possible uses for land taken out of agriculture.

	%	Numbers
Set-aside	41.9	13
Wildlife habitat	3.2	1
Woodland/forestry	12.9	4
Erection of new buildings	22.5	7
Land for other non-agricultural purpose e.g. sports/ recreation	19.3	6
Total	99.8	31

While the numbers involved in "non-agricultural" development during the 1987–91 period were relatively small, it must be remembered that the planning system was acting to constrain the shift of land out of agriculture. Also, many of the tenants in the survey argued that their existing tenancy arrangements precluded alternative land uses, with owner-occupiers having, on the whole, much more freedom of action. The transfer of land during this period (which was undoubtedly a time of great uncertainty in agriculture) affected a quarter of the respondents in the survey. With 13% also contemplating taking land "out of agricultural production", and another 12% "uncertain about doing so", we can see how these marginal proportions begin to add up to considerable possibilities for land development outside agriculture. Further evidence is available from a subsample of 70 farmers who were repeatedly visited for more in-depth, qualitative analysis (see Appendix 1). It was possible to uncover information from this sample on changes in land holding over a 10-year period, from 1980 to 1990. During this period 26 (37%) had expanded their holdings, 34 (48.5%) had seen no change, and 10 (14.5%) had disposed of land. The numbers expanding here were quite large. Clearly, over the longer term the transfers of land were considerably greater.

In order to understand the diversity of farm strategies that underpin these aggregate changes in farming practices and land use, selected case studies are examined in order to show the reasons and processes behind this acquisition and disposal of land. These have been taken from the smaller sample of 70 farms in order to exemplify some of the complex strategies in which land is implicated at the farm level. They are outlined in the next section.

Actor strategies and market arenas

From the perspective of the farm and its principal occupiers – farm families – the uncertainties within the agricultural sector have begun to reflect themselves in capital restructuring of the farm business, rearrangements in the spread and intensity of risk-taking and the reallocation of capital and labour between agriculture and other economic activities (Commins 1990). Under such conditions the degree of commitment to agriculture ("for agriculture's sake") becomes significantly tested, even amongst the largest of farmers where the continuity ethic is central to their business and familial aspirations (see Gasson et al. 1988). In conditions of rapid change and increased uncertainty, the actions and reactions of farm occupiers become much more difficult both to describe and predict. The criteria upon which choices are made and the range of short and long-term options available, are redefined. The degree to which farmers respond to these redefinitions is associated with their relative levels of awareness, rationality, responsiveness to advice, and knowledge from external networks. In particular, the extent to which to which they are tied to sets of internal and external social relations – and the relative irreversibility of these over the medium term – is relevant to any understanding of present and future patterns of action as well as to the assessment of the consequences of those actions. How farmers react to new conditions and trends depends upon the range of options open to them, their evaluation of these options and their ability to pursue a chosen course of action. These factors involve the farm family as a whole and the specific sets of social rules and divisions already established. The interaction between the external influences and the internal structures of the farm necessitates a focus on the external and the internal together, i.e. how farm families are embedded in networks of economic and social relations. In particular the growing problem of debt on working capital becomes a growing external pressure from the mid- 1980s onwards. With the onset of high interest rates in the later part of the decade many indebted farmers were forced into a reappraisal of their farming strategies and their use of assets, often resulting in their participation in new market arenas.

Van der Ploeg (1990) has argued that agricultural practices are necessarily localized, with patterns of "heterogeneity" a recurrent feature of agricultural systems. Variation in farming style is, he argues, too often dismissed as simply a temporary aberration which will eventually be swept away as market efficiency demands a homogenization of agricultural practices. Van der Ploeg demonstrates that an alternative conception of the "market" allows for different interpretations of "heterogeneity". In his view the market is best seen as not only an economic but also as a politically and socially constructed "arena" where "farmers find themselves face to face with state, traders, agribusiness and their advocates" (1990: 267). Within these arenas farmers have differential room for manoeuvre; they have differing resource levels and objectives, and make different calculations as to how these objectives are to be met. This is not to say they are "free agents"; they are caught in *systems of exchange* which seek to bring them into relations of dependency (with input manufacturers and merchants, for instance). According to Van der Ploeg, farmers see this dependency as a "trap to be avoided" (268). This seems to indicate that although farmers are forced to participate in these market arenas one of their aims is simply to maintain as much autonomy as possible. This may not lead to strictly "profit-maximizing" forms of behaviour, but in the long term may offer some protection from the "vagaries" of the market. Such autonomy also allows farmers flexibility in the types of markets open to them. The market "arenas" in which farmers participate can be characterized as "vertical" (that is, within the agricultural commodity sector) or "horizontal" (diversified "rural" economic activities).

Seeing the farmer as an "actor-in-context" is an attempt, therefore, to chart a course between deterministic macro analyses and individualistic attitudinal approaches. However weak farmers may become in political or economic terms, they remain important gatekeepers, overseeing the (re)use of rural land. Land, with its potential for conversion and development, may be a crucial asset in their attempts to preserve autonomy. Thus, the farmer remains an agent of rural change, particularly in relation to the management and use of land, the creation and re-creation of the physical environment. In order to consider the likelihood of certain land-use patterns emerging we need to consider the reactions of farmers to their recent circumstances. We concentrate here on the *use of land* as perhaps the farmer's strongest asset in the struggle to maintain autonomy. As we shall see, farmers have increasingly differing attitudes and objectives in relation to their land holding; these are not simply economic but relate to social and cultural considerations also. These case studies have been chosen from the in-depth farm sample in order to highlight variations in the range of circumstances likely to found under the contemporary conditions of uncertainty.

Agriculture: case studies

Case study 1: farmer A Mr A. farms in north Buckinghamshire and comes from a local farming family. He worked on his father's farm until 1958 when he rented a nearby holding of 60ha. He built up this farm until, in his words, "it was bursting at the seams". He then tried to get the tenancy of a larger farm. Word of this got back to his landlord who offered him a larger holding (an amalgamation of three smaller units) in 1965. When he took over this farm of 118ha it was stocked with pigs, beef and dairy cows and arable land. The pigs were phased out in 1966 and the dairying was subsequently expanded from 30 to 98 cows (dropping back to 80 when quotas were introduced). In 1985 the wheat area was increased and the barley reduced and in 1987 oats were introduced. In 1978 he purchased 15ha "because it was there" from an adjacent farm and a further 22ha in 1986 for the same reason. Part of this land (12ha) is in sole ownership, the rest in joint family ownership. In 1989 he took on the tenancy of 129ha, 48ha of which was put down to set aside by the landlord (Mr A. was contracted to look after this land). The other 80ha was grassland. Mr A. sublet 40ha which he didn't want and expanded the beef enterprise on the rest. He took on the land to maintain his flexibility – "you can always drop it if it gets hard but you can't always take on extra land". In 1991 his tenanted land had increased to 158ha. His strategy was to take on as much land as possible in order to keep his options open. He wanted to expand the livestock enterprises. The terms of his tenancy prevented diversification although he has started to rent his poorest land for horse livery.

Mr A. and his son worked full time on the farm, while his wife worked part time. He employed contractors for spraying, fertilizer application and harvesting. He envisaged using contractors more often in the future.

Annual turnover was between £101,000 and £200,000 and income was around £36,000 per annum. Mr A. had seen his profits slide since 1981 due to "diminished returns from the state, increased costs and the effect of quotas". He had leased in a milk quota but had seen a reduction in the quota from 500,000 to 420,000 litres. His aim was to pay off his overdraft before he retired and allow his son to succeed to the farm.

Mr A. seemed proud of his treatment of the land: "When I first came here it was undrained, wet, weeds, just ticking over. Now it's clean, it's drained and the crops look well. The condition of the soil is improved and no weeds". The farming practices have been "efficient", in response to declining profitability, and there has been investment in machinery (4 tractors, all owned) to allow better cultivation of the land, as well as the use of advisors (ADAS and merchants) on the application of sprays and fertilizers – "I want to leave the farm in better condition than when I took it on".

Mr A. thought that the outlook for farming was continuing decline but saw his own personal situation as good. He felt committed to agriculture and in response to further falls in income he said he would continue with the same pattern of activity but at a lower income level. However, he also indicated that if support prices dropped much further then he would be prepared to sell land for development, put land into set aside and/or intensify production while reducing his unit costs.

Even though he was close to retirement age Mr A. was committed to agriculture, partly because he was trying to ensure that his son succeeded him on the holding. His strategy was to farm efficiently and to maintain sufficient land to keep the enterprise viable. This entailed renting land on a flexible basis around the core which he owned. In fact Mr A. had a pragmatic attitude towards land ownership – "its not important to own your own land for farming purposes, but merely for the pleasure of ownership". It appeared that this farmer had little flexibility in terms of diversification. No strategy of non-farm economic activity was elucidated and it appeared that agriculture was still the centre of his orientation. The use of land, then, was calculated in agricultural terms and maintaining a viable farm business for the purposes of succession seemed to be the overriding aim.

Case study 2: farmer B   Mr B. also comes from a farming family and farms in north Buckinghamshire, close to Milton Keynes. In 1953 he went into partnership with a friend and they became tenants of a 26ha farm. The partnership ended in 1956, although Mr B. continued to farm the land. In 1963 he purchased 40ha on the site he currently farms. Both plots of land – one purchased and one rented – had buildings on them. The owned site had old pig units (former Second World War accommodation blocks) and Mr B. used these for pullet rearing and later for egg production. The rest of the land was given over to corn and sheep. In 1968 a farmhouse was built on the 40ha site. In 1977 he purchased another 34ha. The pullet business was closed in 1979 (the egg production had finished in 1973) because of the cost involved in upgrading the buildings and machinery for this part of the farm enterprise. In 1980 he purchased a further 30ha, so bringing the 34ha site up to 64ha. The sheep and arable enterprises were expanded.

The land Mr B. tenanted was now being incorporated within the Milton Keynes Development Area and being farmed on short-term lets. Twenty hectares of the original farmed area was, by the mid-1980s, being built on and he now farmed 60ha adjacent to it. In 1984 an adjacent farm of 85ha was purchased. He sold the farmhouse and 20ha of the holding to finance the purchase. Here a mixed farming system was adopted because the land was not particularly well suited to arable production.

In 1977 he still employed two men, but one left in 1979 when the pullet business was closed down. Mr B.'s eldest son returned to the farm in 1980 and his second son returned in 1983 (although he left again in 1984 to work on another farm). At the time of interview he still employed one man on the tenanted site but it was thought that man would go when that site was eventually taken over by the Milton Keynes Development Corporation.

This farmer was concerned to develop more diverse economic activities on his farm and began to use his assets for a variety of purposes. The old pig units had been turned into industrial units housing 15 businesses. The conversions were undertaken by Mr B and his son and an agent was employed to deal with the lettings. Mr B. also had a small lake on the farm which was stocked with trout and used for angling. He had also examined the possibility of operating a dry ski slope on the farm (utilizing its proximity to Milton Keynes) but this ran into planning problems associated with access and was eventually ruled out.

In the longer term (five years at least) he wanted both his sons to take over, so one of his main aims was to acquire enough land to allow them to farm successfully. However, Mr B. also had an overdraft which he was keen to reduce. In 1989 he sold 47ha to raise the cash to do this. He felt it important to own land so "you are not beholden to a landlord". In the long term his aim was to buy land to build up the size of the holding for his sons, but in the short term he was forced to sell. Despite selling this land, however, he still remained in debt. He still had a loan on previous land purchases, an overdraft for working capital, and machinery on lease. In 1990, for example, interest and loan repayments absorbed 30% of his income.

Despite these difficulties however, Mr B. had never considered giving up farming and claimed that if the situation deteriorated he would carry on with lower living standards. He was committed to farming and to ensuring that his sons succeeded him.

Mr B. followed a mixed farming pattern, growing mostly wheat, with barley, dried vegetables and oil-seed rape. He had 550 breeding ewes and four tractors (one of which was leased, the others owned). He had developed his land using capital grants for drainage, road building and the construction of farm buildings. He had farmed intensively and, in his view, "efficiently". He was primarily agro-orientated but showed a willingness to diversify where he could. This had usually been in a form which required little labour input (he converted the industrial units but did not manage the businesses in them). Labour time was primarily given over to agriculture.

Mr B. was therefore still firmly established within the agricultural commodity networks and in the face of a squeeze in this sector had attempted to diversify his income sources within the local economy. His longer term strategy, of acquiring enough land to make a viable holding for his sons, had put pressure on the business. He had bought land but during the

1980s found his debt levels rising too fast and was forced to sell land to ease the debt burden. For this farmer land was clearly an asset to acquire and consolidate, but in an emergency could be used to allow capital to be raised in order to solve short term problems. However, he was also willing to use his assets, including land, to diversify his business and thus spread his income sources.

Case study 3: Mr C. After leaving school Mr C. rented land from his father and started farming. In 1973 he managed to get the tenancy of a 36 ha farm. However, he realized after acquiring this tenancy that he would not be able to succeed to his father's (tenanted) farm if he was a tenant elsewhere. He bought his farm in 1979 and then acquired control of his father's farm. After purchase he sold the farmhouse and applied for planning permission to convert a barn for himself. He also sold 10 ha at this time to reduce his debts.

In 1976 he acquired 24 ha of rented summer keep for five years. In 1984 the beef enterprise finished as it was losing cash. At the time he had about 400 animals. He replaced the beef enterprise with sheep (150 ewes). In 1987 he rented 80 ha and in 1988 sold 16 ha to raise money to buy another 11 ha adjoining the other end of the farm. He gave two reasons for this move: first, this was a sandy plot of land which he believed would be suitable for quarrying and, secondly, there was a low wet area on the site which he believed would be suitable for infill.

In the meantime Mr C. lost 60 ha of his rented land as the landlord wanted this for his own son to farm. By 1991 Mr C. found himself short of land and was looking around for another 60 ha to rent. By this time the landfill site (5 ha) was in operation. The rest of the land was for a mixed rotation of wheat, dried vegetables and oil-seed rape. He also had 160 breeding ewes and 40 ha of permanent pasture.

Mr C. described his business practice as "ducking and diving". His land-use strategy was complex and somewhat precarious. While he owned a "core" of 30 ha, the area rented fluctuated from year to year between 80 ha and 160 ha. To maintain this area Mr C. was forced to enter into around eight different short-term tenancy agreements. He referred to the rented land as "insecure" although all of the eight tenancies were for longer than one year, but were not full agricultural tenancies. Mr C. claimed, however, to have trouble renting sufficient land. "I've had trouble getting a right sized unit and trouble renting sufficient land, but I've not enough money to buy. Furthermore, set aside has put a base into the rents" (farmers could set land aside to get the payments for this if rents were lower, so renting land had to compete with set aside).

Mr C. had also made diversification part and parcel of his strategy. As we have seen, he sold and bought land with an eye to diversifying his business. Both the landfill and the sand extraction enterprises were established and Mr C. also undertook a fair amount of agricultural contracting to supplement his income. He was also exploring the possibility of building a restaurant/garage/hotel on a parcel of land adjacent to a proposed new road. As he put it "I've taken a risk buying the low land but I got planning permission for tipping and sand extraction. Its not utopia but it relieves the pressure . . . . Utopia is 50 houses on the front field but we couldn't do that. We've got no woodland, so no war games. If there was a decent living in farming I wouldn't be doing this".

Mr C. had considerable debts. He was still paying off a mortgage on purchased land and had an overdraft for working capital at a time when income had been squeezed – "the viability of the business has gone steadily down since 1981 and the overdraft up". His aim was to get the overdraft reduced and get on a more "even keel" financially. He wanted to acquire more land to extend his arable enterprise and reduce the amount of contracting he was forced to undertake. In terms of farm productivity, Mr C. said "we've had to sharpen the pencil and smarten up our act . . . but we're hard pressed to see how arable productivity could be improved". He saw his land as a "factory floor" and a small conservation area (1 ha) as a "haven". While he wanted to leave his land in better condition than when he received it (he had hopes that his four year old son would succeed him to the farm), he was conscious of the need to exploit modern production methods to raise productivity and took advice from ADAS and his suppliers on the best means to do this. He was, however, aware that he should not "rob" the land.

Mr C. realistically expected family farms such as his own to suffer a lower standard of living in the future. His response would be to continue to find ways to diversify. He believed he was in the fortunate position of having diversification opportunities to exploit and that this was the best way to continue farming. For Mr C. then the land was a means to a variety of ends. His business was taking on a complex shape and he was clearly trying to diversify his sources of income. But this seemed to have been forced upon him. His real aim was to consolidate the agricultural side of the business but his land holding was too small and his debts were too high to allow this to happen. It seemed as though he would be juggling assets and opportunities for some time to come.

These farmers appeared to be very pragmatic in their attitudes to land, using it as an increasingly fluid asset to cope with particular pressures. Farmer A. bought, tenanted, and sublet land. The aim was to take on as much land as possible in order to keep his options open. With a son hoping to follow him onto the farm, he was attempting to build up the core agricultural business. Farmer B. purchased land throughout the 1970s and 1980s and then began to diversify his business in order to raise capital to buy more. However, his levels of borrowing continually frustrated this strategy and he finished up by selling land to ease his debt burden. Again, the long-term aim was to acquire enough land to build up a viable business for his two sons to take over. Farmer C. similarly bought land in the 1970s and then began to diversify in the 1980s, also selling land to pay off debts. He then entered into a series of complex land acquisitions and rental agreements to raise capital and expand the size of the unit. However, once again, debt frustrated his strategy. This suggests there may have been a distinct shift in outlook amongst farmers as they have come to see land as a capital, as opposed to a productive, asset. In this way land is re-commodified around potentially new diversified markets.

Land thus takes its place as one of the assets (along with labour and capital) which become implicated in the farmers' business strategy. We can see from the case studies above that farm strategies in general derive from farmers making calculations about the constraints and opportunities open to them at various junctures. These include not only land-ownership, but policy measures, non-agricultural enterprise choices, need for family continuity, levels of debt and so on. In making decisions these farmers draw upon external sources of advice – on land agents for diversification options and on ADAS for development of agricultural practices, and consult financial institutions, such as banks, over finances. Thus, these farmers are implicated in various networks, some related to policy, others to particular market structures. They must negotiate and "balance" the demands made upon them by their inclusion in these networks. The outcome of these negotiations and "balancing acts" will be a business strategy.

So far we have considered case studies of farmers who have bought and sold land but whose overall aim was to increase the size of the holding to make the farm economically as well as socially sustainable. We now turn to examine farmers whose agricultural land holding was declining.

Case study 4: Mr D. In 1979 Mr D. joined his father and uncle in a partnership and with them purchased the 200ha farm they had previously run as tenants with the help of a 20-year loan from the Agricultural Mortgage Corporation. The partnership was dissolved in 1985 due to financial difficulties and personal difficulties with the uncle. Mr D. then formed a partnership with his father. Dairying had been introduced onto the holding in 1981 and this was intensified after the partnership was dissolved. In order to buy the uncle out, intensify the dairying and generally reduce the level of borrowing, two parcels of land were sold, one in 1981 and the other in 1984 (30ha). This sale allowed a restructuring of the farm's borrowing and was a voluntary decision (in the sense that it was not forced by the bank); as Mr D. put it "we had rising debts but no rising assets". However, further money was borrowed to enable the construction of a new silage pit.

The introduction of quotas had not been particularly traumatic for the farm – "they've changed things a bit but have given a real assured market". By 1991 the owned area remained at 170ha, while 40ha of summer keep was rented and 20ha was rented out. On the agricultural area of 182ha, arable land accounted for 60ha (wheat 35ha and barley 25ha), and permanent pasture 122ha. Mr D. had 75 dairy cows.

Essentially, the farm seemed to be at a standstill in terms of growth. The level of debt and low levels of profitability really gave little room for expansion. There were no plans for diversification and no coherent strategy proposed for achieving growth of the enterprise as a whole. The main enterprise was dairying so the farm was dependent on a sector which is highly regulated by the state. This is a case study of stasis.

Case study 5: Mr E.   Mr E. acquired a farm in 1950, purchased by his mother (he had worked as a farm manager prior to this). He began with one worker milking 22 cows. This lasted until 1961 when the worker left and Mr E. then shifted out of dairying into sheep. He rented 23ha and bought 200 ewes. He also moved to an arable cropping pattern (36ha of wheat/barley). In 1984 he sold 35ha to a neighbour who had "rollover" cash to invest in land. Mr E. took the opportunity to realize this asset. This left him with 5ha owned and 57ha rented. His strategy is to ease out of farming, so he has sublet the rented land. His sale of land cleared all his debts.

Although Mr E.'s son may take on his buildings for conversion to nonagricultural activities he is not interested in succeeding to the farm business. So while Mr E. was coming up to retirement he had no successor and expected to sell his remaining 5ha. He was therefore easing out of farming and in fact his income already came primarily from welfare payments and other unearned sources.

This is therefore a case study of someone retiring from agriculture. The bulk of Mr E.'s land had gone to the neighbouring farmer and his remaining land seemed likely to go the same way. The land was not going out of agriculture but was being incorporated within another holding.

In these two cases we see a slow decline in the agricultural businesses. Farmer D. had no hope of expansion. With high debt levels, and the quota both stabilizing and "freezing" the level of agricultural activity, land was sold to reduce debts. In the last case, Farmer E.'s business was winding down and he was happy to realize his assets. Rising debt levels of debt were thus a major reason for considering either some form of diversification or complete disengagement from agriculture.

142

Conclusion

The case studies illustrate the variety of responses to the recent pressures affecting farmers in the locality. As we have seen, the use of land both as a realizable asset and also as a factor of production is central to the business practices of these farms. However, the strategies of farmers can be traced to a variety of considerations. One of the most important is the question of succession. Where farmers have a clear idea that they will be succeeded they will make arrangements to ensure that a suitable holding is passed on; if they are not to be succeeded they will eventually ease out of farming altogether and will release their land. Crucial to those farms where plans for succession are being made is the size of holding. It may be necessary to expand the size of the holding to make it sustainable in the long run, particularly if more than one child is expected to follow on. Often this long-term strategy can run in opposition to another of the key considerations, the level of debt. As farm incomes have been squeezed, farmers may be forced to find alternative income sources, intensify production from the holding or, failing the success of either of these courses of action, sell land to ease the burden. As the problems faced by farmers have mounted during the 1980s so levels of debt have risen and the use of land for this purpose has become more widespread. Farmers have also recognized the value of keeping a more flexible land holding system. Land is an asset with the potential for diversification, intensification, extension and realization. Where additional land can be purchased then this is generally the preferred option. But keeping a flexible supply of rented land within the holding is also a popular strategy – land can always be disposed of, but not always readily acquired.

The other main strategy is diversification. The case studies show that even those farmers who wish to stay in agriculture are assessing their options in non-agricultural domains. Furthermore, the large scale survey has shown that 54% of farm households in Buckinghamshire were engaged in some off-farm work (Hawkins et al. 1993). Thus, farm households are becoming involved in a variety of market arenas as the pressures on income have forced a recognition of the need to diversify. But for many farmers agricultural intensification is still the more familiar and attractive response to these pressures and they may adopt diversification activities in order to achieve their long-term aim of consolidating the farm business. Many farmers have been active in the agricultural arena for all their working lives and they know its contours intimately. However, this route is, except for the most favoured, becoming harder to take. For the rest it is permanent diversification, disposal of assets and/or (and this was mentioned by a number of respondents) "business as usual" at lower income levels. While this latter choice might be acceptable to more traditional farmers who feel they have nowhere else to go, new entrants or those facing many more years in farming are unlikely to favour a long-term future of "belt tightening". They will seek other solutions.

In terms of land use it is clear that the factors governing the retention in and loss of land from agriculture are complex, based upon different social and economic assessments of the viability of farming and the scope for non-farming enterprises. Partly this will depend on the opportunities offered within other sectors, which are crucially conditioned by the buoyancy of the local/regional economy and the level of planning constraint (see Ch. 9). It also depends on the future of agricultural policy. If farm incomes continue to be squeezed then farmers will seek to ease their debts by selling land. It is likely, however, that the main purchasers of land will still be other farmers – those who are best able to ride out the current recession will become more dominant in the sector. The agricultural "treadmill" grinds on even if fewer farmers are now on it. Nevertheless, as our preceding analysis has shown, the large-scale holdings have not improved their standing over the past 20 years, although the larger of the middle range groups have done so. However, if farm incomes continue to fall and opportunities for diversification do not present themselves then land sales are likely to become more common as farmers try to ease the pressures of debt. Whether this land leaves agriculture, however, is likely to be determined by the effectiveness of the planning system in restraining non-agricultural uses.

To summarize then, the market arenas in which farmers participate are becoming subject to increasing uncertainty. The more diversified world in which farmers now operate confronts them with new levels of risk and new types of negotiation with external agencies, often quite removed from the familiar territory of the food system. Farmers and their families are thus learning fresh "rules of engagement" as they participate in these more unpredictable arenas. Under the old regime of "productivism" land could be regarded as firmly tied to agriculture and this tie stretched back into the past and forward into the future. Now, however, such certainty has gone and land must play its part in the new flexible order. Land is assessed using a variety of criteria, including its agricultural quality, its location and its potential utility in other market arenas. However, there are restrictions on the passage of land from one market to another. These restrictions are codified in the planning system and it is here that further uncertainties and weaknesses in the position of farmers become evident. "Rural" society is clearly no longer "agricultural" society. Many of the villages in Buckinghamshire have become home to a population which has no ties to agriculture and which sees the countryside not as a productive space but as a picturesque setting for their houses and gardens. Thus, the pressures forcing farmers to seek out new uses for their land bring them sharply up against the wishes and aspirations of other rural residents who may actively resist any non-agricultural development in the countryside. In the chapters which follow we document the types of land uses which are deemed acceptable in Aylesbury Vale, and those which are not.

CHAPTER 6

From production to consumption

Introduction

Buckinghamshire farmers are following a variety of strategies which involve the reconstitution of assets, notably land, and diversification of their businesses. In certain instances land use itself can be diversified, notably, as we saw in the previous chapter, into sport and recreation activities. One increasingly prominent use of former agricultural land in the SE is golf courses. Golf is attractive to many landowners as it holds the promise of using land for substantially increased levels of income. Furthermore, as we shall see below, the planning rules surrounding golf course development have been in a state of flux and many planning departments have been uncertain of the proper approach towards these developments. Golf is also interesting because it has traditionally been associated with the middle class and golf courses have been a long-standing feature of suburban and rural landscapes, although there is nothing specifically rural about the sport itself (unlike, for instance, shooting or fishing). Golf is perceived to be "green" in the sense that a golf course retains features that are recognizably "natural" – trees, grass, lakes etc. – however artificial the "whole". It seems to "fit" in the new rural spaces.

During the period of our study there were extensive efforts to produce "value-added countrysides" through the provision of consumption facilities. Developers such as Center Parcs pioneered the construction of partially enclosed rural leisure complexes in attractive landscape areas. Other consumption activities tend to be more extensive landscape users and golf is the most notable of these. In what has been characterized as a "boom", during the late 1980s more than 1,400 planning applications were reputed to have been submitted to planning departments in the UK, involving "an area the size of the Isle of Wight" (*The Guardian*, 15 November 1991). Traditionally, golf courses have taken the form of private courses – the "backbone" of English golf – which account for 80% of the total (Murdoch et al. 1992: 5). Many of the new applications, however, were for commercial courses, either "up-market" country club or hotel-type courses or "down-market" pay-as-you-play type facilities.

146

This sort of boom can be described in the conventional language of economics as the response of market forces to an unmet and growing demand for new golf facilities. We have argued elsewhere (Murdoch & Marsden 1992) that this language renders opaque the ways whereby certain actors attempt to impose new sets of values and uses upon parcels of rural space which had previously embodied quite different sets. In this process, arguments are marshalled concerning the need for diversified economic activities in the countryside and then promoted by developers, *inter alia* linking them to aspects of central government policy and the concerns of other agencies. The countryside is represented as a new type of developmental or consumption space. However, developers must do more than this; they must also show how the proposed development fits into the perceived character of the locality. If there is no perceived "fit" then local representatives may mobilize alternative arguments to show the development to be an unjustifiable intrusion. The provision of golf courses gives us an example of how this process operates in such a way as to deliver types of development, and types of countryside, which meet the demands of certain social groups. As the game has "traditionally been associated with the upper socio-economic sections of the population" (Sports Council South West 1990) and with a third of Britain's 10,000 executives citing golf as their main form of escape from business pressures (*Who's who in industry* 1991), it may be seen as forming a recognizable part of a middle-class lifestyle. We should not therefore be surprised to find golf becoming more securely embedded in the Buckinghamshire countryside.

In this chapter we examine golf course provision in the locality. We show how the provision of this type of consumption good can be differentiated into effectively separate markets, and we examine two case studies. One is at the lower end of the scale, a pay-as-you-play (PAYP) facility, which is closely tied into the locale (both in terms of users and providers). The other is at the top end, an exclusive "country club", where the facility becomes relatively "detached" from the locality. However, before presenting these studies we need to place the game in its regional context.

Golf courses in the South East

We argued in Chapter 5 that the shift away from agricultural productivism in the mid-1980s opened up the space for alternative uses of land. As pressures increased on farm incomes and debt levels rose, farmers and landowners began to search out new sources of income. Moreover, many private and public sector advisors – land agents, MAFF officials, NFU and CLA representatives – began to encourage farmers to look at an increasing range of activities as potential income sources. This shift also coincided with an upsurge in the popularity of leisure pursuits in the countryside (Countryside Commission 1987). One of the most attractive of these new consumption activities was golf. This sport is attractive to landowners (it uses extensive amounts of land, and is therefore a viable replacement for agriculture, requiring many of the land management skills utilized in agricultural production), policy-makers (it takes surplus land out of agricultural production but maintains a notionally "green" countryside), and the many golfers, particularly those un-able to gain access to a course or dissatisfied with the course they use.

We have discussed elsewhere (Murdoch et al. 1992) how the surplus land debate within the agricultural community translated itself into a set of issues concerned with unmet demand within the golfing sector. Reports such as those emanating from the Country Landowners Association (CLA 1988) and the Royal & Ancient Club of St Andrews (R&A 1989) raised the spectre of a "hidden" army of frustrated golfers who were denied access to courses. The former, working on the principle that there should be a national average of one golf course per 25,000 of the population, argued for 341 new golf courses while the latter called for 700 new courses by the year 2000. The R&A report cited as evidence for under-provision the existence of excessive demand, such as the pressure upon membership waiting lists in the country as a whole (waits of three to five years were not exceptional). The problem was seen to vary according to region but it was generally accepted that the SE was particularly poorly served. Within the region demand was acute due both to the popularity of the game and the numbers of middle-class residents. In a Sports Council survey of courses in the south published in 1989, 59 of the 64 clubs that responded to the questionnaire reported waiting lists for full membership. The Council concluded that: "Most golf courses in the Southern Region are operating at or just below their capacity. Whether this capacity is imposed by a club committee wishing to provide pleasant playing conditions for its members or by the absolute limits of the course's ability to accommodate a certain number of rounds, the end result is the same: the casual but potentially regular non-club golfer is restricted in his [sic] efforts to pursue his sport". This conclusion reiterates the more general argument put forward by the Sports Council, the R&A, and other sporting and landowning groups, that there was considerable untapped demand in the golfing sector, particularly at the bottom end of the market where "casual" golfers were finding it difficult to play at regular venues. The problem was seen as particularly acute in certain areas within the Southern Region: "Oxfordshire, central Buckinghamshire, western Berkshire, central and southwestern Hampshire are all poorly provided with pay as you play courses and therefore the residents in these areas have to travel substantial distances to play the game unless they are members of local private clubs" (Sports Council 1989: 7).

The "boom" in golf course applications during the late 1980s was concentrated on these counties, particularly on the "underprovided" counties of Oxfordshire and Buckinghamshire. Although the majority of applications were of the 18-hole private club variety some facilities were aimed at the lower end of the market where PAYP courses are provided to serve those starting out in the game. In this chapter we examine the provision of two contrasting facilities in Aylesbury Vale and attempt to draw out some of the implications for the "post-productivist" countryside.

Golf course provision in Aylesbury Vale

As we indicated above, the new leisure uses of rural land derive from two main sources; first, the increased amount of land deemed to be surplus to agricultural requirements and, secondly, the demands made on this land by mobile urban residents or ex-urban rural residents. These trends were explicitly recognized in the leisure strategy document produced by the County Council in 1989 which noted that 2,496 ha of land in Buckinghamshire had been entered into the set aside programme, thus "there is a new opportunity to produce a more diverse landscape than in the past" (para. 1). Moreover, "there is evidence that a more affluent and mobile population, with more leisure time, is looking for an increasing range of recreation opportunities in the countryside" (para. 2). The strategy also noted that the Sports Council had identified the need for further golf courses in the Aylesbury Vale and High Wycombe areas, particularly in the PAYP category. Furthermore "new courses are also needed to serve the northern part of the county in response to the expected growth in population by the end of the century" (para. 112). The new courses, it was argued, should follow certain guidelines, including that they: (a) be sited on land of lower agricultural, landscape and ecological value; (b) be close to urban areas where demand is concentrated; (c) have special regard to the conservation of the character of the landscape and wildlife; (d) have a good, safe road access; (e) incorporate public access and (f) include provision for the creation of features of nature conservation and landscape interest.

At this time there were 24 golf courses in Buckinghamshire, four of which were public (two in Milton Keynes and two in the south of the county). The majority (16) of these courses were in the southern districts and five were in Aylesbury Vale. The latter were either south of Aylesbury or around Milton Keynes and Buckingham. There were no courses in the middle of the district (see Fig. 6.1).

When the golf course boom began to take off, Aylesbury Vale, situated in an affluent corner of the region, was a prime target for development. Between 1987 and the end of 1992 AVDC received 24 applications for golfing facilities. Of these, six were refused permission (three golf courses

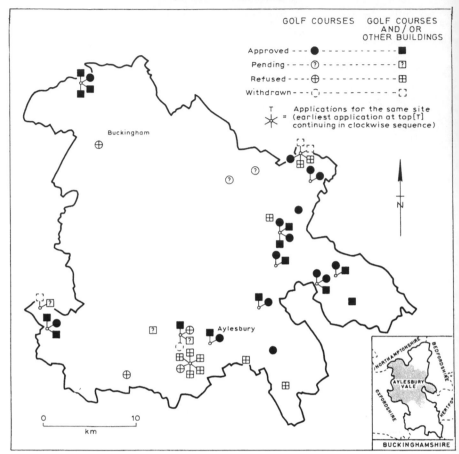

Figure 6.1 Golf course applications, Aylesbury Vale, 1987–91.

and three driving ranges) and 17 were given approval (including one course extension) (see Fig. 6.1). Fourteen of these were for golf courses and five for driving ranges (two courses had attached driving ranges). Six of these gained permission for accompanying buildings. Had all these new permissions been acted upon the number of courses would have gone from five to 21, a huge increase. However, at the time of writing only four of these new courses had gone into the construction phase, partly due to the downturn in the regional economy.

According to one respondent, well placed within the District Council, the planning department was caught without a clear policy when the "boom" struck. He claimed the Council had to abide by the advice they were receiving from six sources, all of which were largely permissive of golf course development. These sources of advice included: the DOE's Planning Policy Guidance Note (PPG) 1, which said development should be regulated by the market place; contemporary public concerns about surplus agricultural land; MAFF support for "reversible development" which initially included golf courses, meaning that MAFF would only object to uses of Grade 1 land (a position not modified until the revised version of PPG 7 in January 1992); advice from the Council's Leisure Services Officer, who believed that Aylesbury Vale was hopelessly underprovided with golf courses; the experience of appeals in the south of the county which had been lost by Wycombe District Council on the grounds that there was no material change to the countryside with golf development; the English Golf Union and land agents such as Savilles and Humberts, who proposed that demand for these facilities was high. For all these reasons the District Council felt pressurized into granting permissions.

Nevertheless, by 1991 AVDC had begun to tighten up its policy. It was being extremely careful in allowing any ancillary development "riding on the back" of courses and there was a general feeling that enough golf course permissions had been given. In the *Rural Areas Local Plan* of 1991 a much clearer exposition of the Council's policy was given. It was argued that "the provision of further golf courses in the District has to take into account the community's needs and demands for this type of recreational activity and the need to protect the open countryside" (para. 9.51). On "need" the construction of three new courses had brought the District below the Sports Council's provision threshold of one nine-hole course per 12,000 to 15,000 of population (the District figure stood at one course per 10,000 of population). However, these standards do not take account of the geographical distribution of the courses and the Plan noted that courses in Aylesbury Vale were concentrated in the eastern part of the District adjacent to the main centres of population. There were far fewer courses in the central and western parts but here population densities were much lower. In future, it was argued: "Consideration of proposals for golf courses should . . . take account of the existing provision of such facilities, including those under construction and those committed, and the need to be well related to the centres of population they are to serve" (para. 9.52). Despite this overall fulfilment of requirements, the Plan also argued that "there is an identified need to increase access for the casual golfer to pay-as-you-play facilities" (para. 9.53). This "need" had been identified in a report by L&R Leisure Group for AVDC in October 1988: "A public pay-as-you-play golf facility is required within the Aylesbury area. The pressure on the majority of private golf clubs in the area combined with a shortfall in provision in comparison with Southern Region Sports Council recommendations indicates there is frustrated demand for this type of golfing facility" (12).

The *Rural Areas Local Plan* recognized this "need" and indicated that "one course serving the Aylesbury area, is planned, but not yet started as part of the Coldharbour Farm development. A further pay-as-you-play facility is recommended by the Sports Council, ideally sited near Buckingham" (para. 9.53). The Coldharbour Farm development was an area of 217ha just to the west of Aylesbury on the outskirts of the town and was indicated in the *Aylesbury Planning Study* of 1981 as having potential for the accommodation of residential development. The *Aylesbury Local Plan* of 1988 confirmed this potential and allocated the land for a mixture of residential, employment, recreational, educational and community uses. A golf course was proposed as part of the recreational strategy:

> Although there are several privately controlled full-sized golf courses not far from Aylesbury, there are none in close proximity to the town. The nearest facility is the 9-hole course at Weston Turville. The Sports Council's recommendation of one 18-hole course for a catchment of 20 to 30 thousand people suggests that provision of at least one close to the town is justified. This would be appropriately located in conjunction with other recreational facilities on the periphery of the town. (*Coldharbour Farm Planning Brief* AVDC 1988 para. 8.33)

Although this land was proposed for development the scheme was dependent on outside development companies getting involved and the onset of recession in the late 1980s seemed to have reduced their interest.

As this account indicates, planners prevaricated in their attitude to new golf courses and for a while it was relatively easy to get planning permission to convert farmland to this new use. However, planning permission is only the first hurdle in the golf course development process; for courses to be built, markets must be identified, capital must be raised and business strategies must be credible. These elements are not always readily available and many planning permissions are simply not converted into actual courses. However, in the Vale a few new courses have gone into the construction phase and in the next section we look at two examples.

Golf: the case studies

The perceived need for a PAYP facility in the Aylesbury area led to another proposal coming forward which we shall study in some detail here. This initially involved the development of a driving range and golf centre and then subsequently a 9-hole golf course. The site for this development was a farm close to Aylesbury. The farmer had been in the locality for most of his life, taking over the farm in the immediate post-war period. The land was a mixture of tenancy and freehold ownership. The farm enterprises were exclusively livestock production and it was dependence on these sources of income that forced the farmer to seek out alternative economic activities. A fall in livestock prices in the latter half of the 1980s put pressure on farm income and led to the search for diversification.

The first of these alternative enterprises was the conversion of a farmworker's cottage into a holiday home. This was successfully achieved and the cottage was used almost all year round. However, the enterprise was the domain of the farmer's wife and was, in any case, incapable of delivering the required income levels. Pressure was intensified by increases in interest rates which pushed up the farmer's overdraft. He was forced to sell a parcel of land. At this time he began to search around for another enterprise and he approached a planner–surveyor to seek some assistance in identifying a suitable activity for the farm, stipulating that it should be recreational with no noise or smell ("I'm carrying on living here so what comes will affect me"). He also took advice from the District Council's Leisure Services Department. His consultant believed he had identified resources on the farm which could meet a slot in the market: a field and a barn could be adapted to a driving range. After discussion it was agreed that the consultant should draw up a planning application.

Again the District Council Leisure Services adviser (who was, incidentally, a golfer) was consulted about the application. He told the developer that the local authority would probably be able to "buy" time on the range in order to teach people to play golf as part of the District's leisure strategy. He also gave advice on management, equipment, insurance, etc. Further, he initiated a meeting with another golf centre owner for business advice (sources of capital etc.). It was estimated that the business would cost around £100,000 to establish, with an estimated pay-back time of five to seven years. In order to raise the required capital a partnership was established between the farmer/landowner and the planning consultant, plus the latter's brother. It was agreed that the farmer would contribute the land and the building (around £45,000 worth of the cost) and the other partners would put up the capital and the "expertise". In the farmer's view, the enterprise "will suit the farm". First, labour from the farm business could be "contracted out" to the golf centre for such tasks as golf ball collection, earth moving, etc. Secondly, at the time the project was initiated official MAFF policy was that golf course development was reversible and the same principle applied to golf ranges. This allowed the farmer to take land out of agriculture and to maintain his farm buildings, particularly a 200-year old barn, which would be part of the driving range. The farmer also argued that the scheme would not damage the environment: "I'm not an environmental fanatic but what we propose will not damage wildlife. The site will be screened and bird life won't be affected".

The response of local residents to the plan was mixed. According to the developer: "Local individuals are for it but the Parish Council has expressed concern". The concern of the Parish Council was based on four issues: first, the range constituted development outside the village boundary and was therefore an incursion into open countryside; second, the floodlighting would disturb nearby homes; thirdly, there would be a problem with access from the main road and the levels of traffic; and fourthly, it would set a precedent for development outside the village, threatening in particular the division between Aylesbury and the village. The farmer believed these fears to be exaggerated. In response to concerns about traffic he agreed not to open the range until 10 am and gained agreement from the Highways Department that the road would be appropriate. On the floodlighting he believed that "they don't understand the impact the floodlighting will have. They are thinking of floodlights on a football pitch but these will be at a much lower level". The Parish Council's arguments did not forestall the development and it received planning permission.

The idea of having an adjoining golf course came onto the agenda shortly after the range opened in June 1991. It was the golf centre manager who raised the issue. He argued that a 9-hole golf course would provide the bottom rung on the golf ladder for those golfers without a handicap who were excluded from other (private) courses. A parcel of land adjacent to the range was seen as a "natural" site for the course. Again, the farmer put in the land (15 ha of Grade 4 land) and the other partners put up part of the capital (the rest was provided from a loan). The course was designed by the two partners (who were golfers) and the golf centre manager.

There was more vocal opposition in the area to this second proposal. A meeting was held in the village and was attended by 13 villagers expressing "united opposition" to the course. One Parish Councillor said: "This is going to affect [the village] to a large degree, and we cannot close our eyes to it" (*Bucks Herald*, 31 January 1991). Again, the Parish Council objected to the proposal on the following grounds: "(1) The site lies outside the village limits for development; (2) the proposal would result in a significant increase in turning traffic on the already busy A418 affecting residents of nearby dwellings; (3) the proposal is to use open countryside on a commercial basis to which the Parish Council are opposed; and (4) there is inadequate access and car parking to serve the site" (AVDC Planning File). Five further letters were received by the Planning Office expressing support for the proposal mainly on the grounds that the area needed this type of amenity centre.

In evaluating the proposal the Planning Office took the following criteria into account: "Recent Government advice has indicated that there is a surplus of food production and has encouraged the diversification of the rural economy particularly where this involves lower grade agricultural land so as to open up wider and more varied employment opportunities. That need however is to be balanced against the protection of the environment without giving agricultural production a special priority". However, it was recognized that: "The Southern Council for Sport has identified a need for more golfing facilities in the Aylesbury and High Wycombe area (particularly "pay as you play" type facilities)". Furthermore: "The [*Rural Areas Local Plan*] indicates that proposals for golf courses in open countryside will be sympathetically considered subject to certain criteria and subject to a requirement that if new building is required it should be kept to a minimum and should be suitably located and landscaped to mitigate any damaging effects on the character and appearance of the area". It was further argued that golf courses may improve "the character of the countryside emphasizing natural features or providing additional landscaping or tree planting" (AVDC Planning File). The planners recommended approval of this development and permission was granted, subject to certain conditions, including an agreement to improve access and a screening plantation of trees.

The work on the course was undertaken by the "farm" side of the business; as the farmer pointed out: "I've made a living from grass all my life – as a farmer I am a golf course specialist". Special turf was purchased for the tees and greens and some mounding was provided by spoil from the range (which had been clamped) with topsoil taken from the widening of the car park. The land on which the course was built had been down to silage from the previous year, so once the grass was taken off in the early summer it went straight into golf.

The financing of the course was calculated on a ten year programme. Investment in the whole enterprise amounted to £400,000 and it was anticipated that after 10 years the centre would be earning over £40,000 per year. Already the driving range was doing better than they had projected. It had been expected that the maximum numbers using the range in any one week would be 1,050 during the summer months, falling to half that in the winter. In actuality the numbers using the range were 10% above these projections. The course was expected to attract 620 people per week in the summer months, declining to less than half this figure in the winter. The green fees would be £8 during the week and £9 at weekends. There was also a membership scheme being introduced costing £50 per year for which £3 was deducted from the usual green fees and members had privileged access as well as the opportunity to play in competitions. Clearly this was to be a down-market course aimed at those starting out on their golfing "careers". As the farmer put it: "We're looking for a particular segment of the golfing population – beginners, average golfers – so we don't want the course too difficult. We get skilled golfers on the range – members of other clubs – but they won't regularly play on the course". By mid-1992, even before the course was completed, 230 people had joined the club (the target was 250 men and 50 women!).

A greenkeeper was appointed and one more part-time worker would be needed once the course was up and running. The policy was to recruit labour locally. The manager, the chef and the two people working in the shop all lived locally. The farmer reasoned as follows: "My agent has lived in the District for a long while, I've lived here a long while. He lives in the next village and his partner also nearby. Our local reputation is at stake as we all live locally". This enterprise was "local"; it was developed and managed by local agents to serve local consumers. As the golf centre was aimed at the "bottom" end of the market it necessarily relied on the patronage of golfers from nearby towns. We can contrast this with another course which was established at roughly the same time and which is much more "detached" from the locality, both in terms of its development and use.

This latter type of golfing facility was to be found in the grounds of a large country house to the east of Aylesbury. Here a farmer with a 6,250ha holding (4,583 owned, 1667 rented) had decided at the end of the 1980s to develop an alternative economic activity. This was an intensive arable farm growing mainly barley, wheat, and dried vegetables. The farm had an annual turnover of around £500,000 and had been run with an overdraft since 1976. The pressures on the farming sector in the late 1980s forced the search for another enterprise. Initially the farmer looked at horses as a source of income but the land was not suitable. In conjunction with a consultant, a plan for an 18-hole golf course was then drawn up and gained planning permission in 1989. Unlike the previous case study this farmer did not intend to develop the course himself; he wished to leave the construction and administration of the club to a leisure company. Although planning permission was secured at this time, it took another 12 months before a developer could be found to construct the course. Even then the firm would only consider the venture if a further 18 holes could be added. In their view only a 36-hole course would be profitable. The farmer effectively leased the land to the development company although he still remained involved in the project and the farm business undertook some of the construction work. The development company itself was based in London and specialized in leisure developments. It in turn subcontracted aspects of design and construction to other smaller specialist companies. According to the farmer, this venture was undertaken at just the right time as they managed to achieve permission for the 36-hole course and clubhouse without too much difficulty.

The course itself was to be set in the grounds of the country house and this provided a picturesque parkland setting – a far cry from the stark agricultural land of our earlier case – and was marketed as a "Golf and Country Club", with not only a 36-hole course but swimming pool, sauna, jacuzzi and other facilities. Membership was organized on a debenture basis whereby the prospective members become "shareholders" in the club. These shares were to be tradeable (but only after the club had reached its maximum membership total, which in this case was 900 – at the time of writing, one year after the club opened, 500 had been sold). The debentures were divided into various categories: ordinary debentures costing £3,600 entailing the payment of an annual subscription fee (£600): and lifetime debentures of £10,000 which had no annual fee. A husband and wife scheme was also available for £5,700 per couple. There were, however, no formal restrictions on membership, such as possession of a handicap, other than ability to pay the required debenture fee. It is clear, however, that such fees would be out of range of all but the wealthy. This course was aimed at the more "exclusive" end of the golfing market.

These two case studies exemplify two contrasting forms of golf course provision. The first was a cheap facility, mainly serving the immediate locality. It aimed to provide the first step on the golfing ladder for in-experienced players, while giving members of other clubs the chance to practise on the driving range. The club emphasized its rôle in the locality; the developer stressed his affiliation to the local community and high-lighted the contribution the course would make. The second case was a course which fulfilled a quite different function. It was an up-market, expensive and therefore exclusive course catering for a golfing popula-tion with the financial means necessary to purchase the social *cachet* asso-ciated with membership of such a club. Its links with the locality were tenuous; it advertised for members in London as well as locally and served a golf market extending much further afield than that of the first case study. It was in effect relatively detached from the locality.

These two cases show how rural land can be re-commodified; how it can be placed in quite different markets even within the same broad sec-tor. Clearly, these markets are limited; there can only be so many PAYP golf courses and up-market country clubs in Aylesbury Vale and we have examined two case studies in which the landowners and developers made the shift at the right time. However, these new land uses mark a change in the complexion of the Buckinghamshire countryside and show how land moving into new activities comes to enshrine new sets of social relations.

Conclusion

As farmers seek to diversify their enterprises and find new uses for their land they become involved in new economic domains in which they may have little expertise. The market arenas in which they find themselves are complex and are often surrounded by, or constituted through, distinctive regulatory arrangements. Under these circumstances farmers' responses may vary. Our two golf case studies reflect this. The first case clearly demonstrates how a farmer/landowner entered a new market arena and used his local social and political networks (e.g. he was well known in the locality and a member of the District Council) and the resources (including, crucially, his local knowledge) these bestowed on him, both to establish and expand a new business. This was by its very nature (low cost and accessible), embedded in the immediate locale. He played to his strengths. The second case study was of a rather different type. Here a farmer/landowner with extensive material resources (i.e. land) overcame the problems of uncertainty posed by engagement with a new market arena by building a network with a non-local development company which (through its ownership of capital) effectively dictated the terms of development. While the farmer/landowner increased his income he ceded control of the golfing business (and was left to concentrate on the enterprise he knew best – farming). This business was therefore a much less local concern (the locality merely provided a picturesque backdrop). It was expensive and sought out its customers from a wide catchment area. The "positional status" of the club was marked out by the size of the membership/debenture fee.

What this shows is the way various actors come together to promote new uses and markets for rural space. Land is accorded a new rôle and so are the farmers-cum-developers-cum-golf operators. New representations of rural space are also mustered to facilitate this change of use. The process of conversion documented here was initiated by farmers/landowners keen to shift land out of agricultural production. Golf was an attractive option because, as the farmer in the first case study put it, "it will suit the farm". Golf allows a new market sector to be opened while at the same time using many of the farm's existing resources – land, labour, machinery and management know-how. What are required are new sources of capital, and the pursuit of these often entails farmers either going into partnership with others or ceding control of the new business to a development company.

From the perspective of the planners, the perceived "need" to allow land to be shifted out of agriculture was bolstered by talk of "untapped demand" in the golfing sector. This discourse came from golfing and farming bodies (see Murdoch et al. 1992) and was increasingly influential in swaying the decisions of planning authorities, particularly where they had no clear policy on golf courses. Furthermore, proponents of increased levels of golfing facilities in the countryside were not just talking about aggregate demand, for there was an increasing appreciation of the complexity of the golfing market. As the Country Landowners Association, for instance, made clear:

"Golfers can be divided into a number of market sectors which conform to a greater or lesser degree to the overall profile of the golfing population which is predominantly male, managerial or professional and car-owning. The section that can be identified are regular players, frequent players, club players, novices, casual players, tourists, business players and prospective players. Each have their own requirements in terms of the sophistication of facilities offered and the success of new development requires consideration of what market or combination of markets is to be served" (1988: 1–3).

The two cases examined here clearly aimed for different market sectors. The PAYP course would capture "casual players", "novices", and the "prospective players" (although regular club golfers use the driving range). The country club would appeal to "regular", "frequent" and "business" players. Thus, these two new parcels of countryside become new types of social space, accessible to certain social groups. The PAYP course resembles much more the municipal course and is (for a small fee) effectively open to all. The country club on the other hand provides a luxurious and highly desirable consumption environment and is expensive, thereby ensuring social selectivity. The new uses and values placed upon this land (re)produce certain representations of the rural.

From the extensive social survey of Buckinghamshire residents it would appear that there is some limited support for golf courses in the countryside, with 13.7% of the sample wishing to see more and 43.8% happy with the existing provision. However, 42.4% wanted to see less courses (for the seven villages these figures were much the same with 16.5% wishing to see more courses, 41.5% the same amount, and 42% wanting less golf in the countryside). Coming in the wake of a pronounced increase in golf course planning permissions, these results would seem to indicate that if the majority of the new proposals are constructed there will be little support in the locality for this new form of "rurality".

CHAPTER 7

Minerals development in a hostile environment

Introduction

In the previous chapter we examined an example of land development which proved attractive to farmers, landowners and those seeking recreational opportunities in the countryside. Furthermore, golf course development was also not fully "captured" by the planning system and, for a period (now probably over), there was a rush of golf course development proposals, the majority of which gained planning permission. We now turn in the next two chapters to cases of land development which may be attractive to landowners and developers but do not fit at all easily into the types of countryside sought by the middle class. Furthermore, these are traditional uses for rural land and are therefore surrounded by much more elaborate planning frameworks than those which exist for golf. Planning plays a dual rôle in these two case studies, for it not only imposes the prevention of development in the countryside through development control decisions, but also attempts to ensure that provision is made to allow these traditional land uses to continue. As we shall see, planning authorities can run into considerable difficulty in achieving the delicate "balancing act" that this dual rôle demands, particularly where influential groups in the locality are effectively orchestrating opposition to these types of development.

We begin with minerals, an established and traditional user of land, but one which often jars with many people's conceptions of legitimate uses of the countryside. Minerals development has also become something of a pressing issue in debates about the use of rural space as economic growth, mediated through levels of house- and road-building, has progressively increased demand for quarry products such as crushed rock and sand and gravel. In what follows we concentrate on sand and gravel exploitation. This is the most common mineral in the study area and is thus something of a contentious issue for local residents, as we shall see below. Furthermore, the close proximity of Aylesbury Vale to London means that there has been an almost continuous demand for minerals in Buckinghamshire. Sand and gravel also has its own distinctive

160

planning regime which provides a "window" on the minerals development process. Before we turn to our local case study, therefore, we must spend some time considering how the economic context is shaped by the political/planning framework. In this sector an elaborate planning structure is in place, extending from the "national" to the "local" which surrounds, and to a certain extent defines, the "market". Necessarily therefore, the context here is complex and requires a detailed exposition. Following the case study we return to this contextual framework to show how the reverberations from the "local" development process could be heard in a "national" debate over minerals planning.

As we emphasized in Chapter 2, the SE has been subject to a substantial rate of economic growth for much of the past 20 years. Economic growth has been accompanied by population growth, leading to intense periods of house- and road-building and other infrastructure provision. The construction industry experienced boom and slump during this period, with the good times coming in the late 1960s, early 1970s, and most recently in the latter half of the 1980s. At the present time we are struggling to come out of a recession which has hit the construction industry particularly hard. As approximately 90% of all aggregates are used by this industry the production of minerals is to some considerable extent bound up with this sector. Total consumption of all aggregates reached its peak in the SE in 1973 when it was estimated to be 66 million tonnes (mt), thereafter falling to 48.3 mt in 1977 before rising again to 55 mt in 1985 (Bucks County Council 1989b). However, this figure was surpassed in 1989 when it reached 74.4 mt (DoE 1991b).

Buckinghamshire is at the heartland of growth in the region. It has therefore generated its own level of demand for aggregates. In the mid-1960s the county consumed nearly 80% of the aggregates it produced, this figure falling to 42% in 1970. By the mid-1970s however, Buckinghamshire was consuming more than it produced, a situation that has prevailed ever since (Bucks County Council 1978). The main reason for this was the development of Milton Keynes. However, such figures tell us little about the pressures on mineral-bearing land in the county as mineral demand is calculated on a regional basis. In any case it was stipulated in the Act permitting the development of Milton Keynes that the minerals used in its construction should come mainly from outside the region and they do indeed come from Northamptonshire and Bedfordshire. Northwest Buckinghamshire and Aylesbury receive some of their requirements from Oxfordshire, though how much is unknown (Bucks County Council 1978).

The main pressure on Buckinghamshire comes not from internal demand but from London. The capital has traditionally obtained the majority of its supplies from within a 30-mile radius which includes the south of the county. Minerals extracted in the south of Buckinghamshire have been used mainly in London but also in Slough, High Wycombe, Marlow, Maidenhead and Reading (Bucks County Council 1978). It is this pressure in the south, where most of the mineral-bearing land is to be found, that has raised the political stakes in minerals planning as we shall see below. Within the region Buckinghamshire is not a major producer of sand and gravel (at 1.0 to 1.5 mt per annum compared to nearly 4 mt in Surrey) and currently there are approximately 40 quarries with around 25 operators.

The amount of social change taking place in Buckinghamshire during the past 30 years has heightened opposition to mineral working. There was some evidence of this in our social survey where only 4% of the total sample wanted more quarries in the countryside while 54.3% wanted less (in the seven villages the figures were 3.6% and 54% respectively). The majority in favour of less minerals working was overwhelming, with very few residents wishing to see any increase in the amount of extraction. This attitude among rural residents clearly poses a threat to the location of new mineral quarries in the Buckinghamshire countryside. However, this state of affairs is by no means new and the planning authorities have taken steps to confront the problem posed by increased demand for aggregates in the face of public hostility.

The minerals planning context

Separate structures and procedures for minerals planning have evolved because of the distinctive characteristics of the minerals land-development process which sets the sector apart from the generality of land-use planning. This is evident in the following respects: (a) minerals and waste disposal are the only subjects where development planning and planning development are the sole responsibility of county councils; (b) minerals planning has its own distinct body of legislation and policy, codified at the local level in minerals plans and at the national level in Minerals Policy Guidance Notes (MPGs). (This is the only industry to receive such formal, separate policy treatment within the planning system); and (c) minerals planning differs significantly from more general land-use planning in the following respects: requirements to avoid sterilizing minerals resources; the validity (including time limit) of permissions; the conditions which can be attached to permissions; and conditions covering aftercare and restoration of sites (see Lowe et al. 1993). A standard explanation of this regime refers to the peculiar nature of minerals working as a form of development. Minerals working is not *development* of land but its *exploitation;* it is a disruptive activity which renders the land less valuable than before. However, minerals can only be extracted where they lie and this gives a crucial rôle to the planning system to secure the necessary conditions for exploitation to take place. If demand for minerals is high then there will be pressure on planners to allow minerals to be quarried. However, local resistance to exploitation may also be acute, thus placing planners in the invidious position of balancing one set of (national) concerns against other (localized) ones.

The combination of economic pressure for minerals and social resistance to the exploitation of land in residential areas has given minerals planning in Buckinghamshire a particularly high profile. The pressure on local politicians and professional planners has manifested itself in the shape of numerous attempts by these authorities to forge a consensus out the minerals planning system. Such attempts began when the County Council was confirmed as the Minerals Planning Authority in the 1971 Minerals Planning Act. Conscious of the competing pressures upon them, the County Council began work on a minerals plan for South Buckinghamshire. In the 1974 plan *Gravel Working in South Bucks* specific sites for gravel extraction – so-called "preferred areas"- were identified. This was by no means a comprehensive analysis of possible sites. In fact all the sites identified in the Plan were in one area – the Colne Valley.

The suitability of the identified sites was almost immediately challenged by a local company, Hoveringham Gravels, in a planning appeal about land at All Souls Farm, Wexham. Although the appeal was rejected, the Secretary of State questioned the status of the Plan when, in his decision letter, he argued that the provision to meet demand in the area was "subject to severe qualifications and is quite likely to prove seriously deficient for that purpose" (quoted in Bucks County Council 1978: para. 52). The County Council took this to imply that they were in danger of losing subsequent appeals unless they firmed up the minerals plan, so they began work on a comprehensive county-wide study of minerals resources to see if "permitted reserves" could be increased.

This first plan was drawn up between 1975 and 1978 when it was put out for public consultation. In the meantime the Verney Committee, set up by the Government to investigate the whole issue of planning for aggregates, had reported in 1975. The County Council sought to ensure that the Committee's recommendations were incorporated in the plan. Verney had proposed: "(i) that planning authorities in areas of high demand should reappraise the priority accorded to aggregates production with a view to ensuring continuation of locally produced supplies at recent levels for the next 10–15 years; and (ii) the Standing Conference on London and South East Regional Planning (SERPLAN) should give more precise guidance to its constituent bodies on release of land for aggregates working" (quoted in Bucks County Council 1978: para. 33). The latter proposal effectively pointed to the need for much more centralized direction of local minerals planning. This would be undertaken through the Regional Aggregates Working Parties (RAWPS) of which there were ten, covering the UK. These working parties facilitated close links between the industry and the planning authorities and their membership was made up of County Planning Officers, representatives of the minerals industry and DoE personnel. The work of the committees was overseen by a National Coordinating Group (NCG), chaired by the DoE, composed of working party chairmen and representatives of the industry. Although this committee structure gave the industry good representation at the strategic planning level, Verney's first proposal effectively limited the amount of aggregate working in the SE to that of "recent levels" (i.e. based on the post–1973 slump). This was accepted by Buckinghamshire County Council in their 1978 Draft Plan which stated that "land should be released to sustain levels approximating to recent production rates and that for the next 10 years this could mean gravel extraction yielding up to 16 mt in the county" (Bucks County Council 1978: para. 58).

With the overall production quota being supplied by the Verney Committee it only remained for the County Council to determine where minerals should be excavated in the county. This was, however, by no means straightforward for "a proper balance has to be struck between meeting the need for aggregates and protecting the environment" (Bucks County Council 1978: para. 65). What the planners needed to undertake this exercise was some tool which allowed them to achieve the implied political balancing act. Such a tool had been provided in a 1976 DoE consultation paper suggesting the use of the "sieve" map approach. This technique, the Council believed, would allow for "a systematic and rigorous analysis of the aggregate bearing areas of the county" leading to sites being identified "as objectively as possible" (Bucks County Council 1978: para. 65).

The first stage in the use of the "sieve map" involved surveying the extent of total known sand and gravel deposits in the county. It was here that the first problems arose for there was simply not enough geological information to do this in north and mid-Buckinghamshire. The second stage involved taking out those areas which were subject to constraints (such as grades 1, 2 and 3a agricultural land, nature conservation sites, archaeological sites and, subject to certain caveats, AONBs and Areas of Attractive Landscape [AALs]). However because of the gaps in the geological information this could only be fully done in the south.

By using a seemingly objective procedure to identify accessible minerals reserves, the planners hoped to distance themselves from and defuse future political controversy over particular sites: "By using a sieve map approach all the factors that have been used to refuse planning applications for mineral working have been explicitly stated. The aim is to demonstrate the rational way in which the preferred areas have emerged and to avoid creating the sort of uncertainty that has been experienced by the public and the operators in the past" (Bucks County Council 1978: para. 69).

This attempt to mould a consensus behind the plan using the very visible, yet "objective" and "rational", sieve map was not successful. First, because of the paucity of geological information in the north all the "preferred areas" were in the south (although this clearly "left the door open" for minerals companies to find sites in the north). It was stated in the ensuing Plan that "the difference in presentation between potential gravel working areas in south Buckinghamshire and central and north Buckinghamshire does not reflect a variation in emphasis but merely one of technical feasibility" (Appendix 4). However, such an approach was hardly likely to reassure those residents of south Bucks who already felt that they had taken an unfair amount of mineral working in their area. Secondly, the use of the sieve map did not please certain members of the minerals industry, as it enabled the Council to identify areas "where there would be a presumption against working aggregates, to be ruled out" (Bucks County Council 1978: para. 65). In fact, a challenge to the whole exercise was subsequently mounted by the Sand and Gravel Association (SAGA) in the High Court.

The Minerals Plan, having been through its consultation period, went to a Public Inquiry in October 1980 and this lasted for three months. The Inspector reported in August 1981 and the Council, having made all the modifications he suggested, adopted the Plan in February 1982. It became operative in April of that year. At this point SAGA mounted their High Court challenge (prompted by one of their member companies, Halls Aggregates, who owned a site outside the "preferred areas"). SAGA objected to the Plan on a number of grounds which can be grouped under two headings; procedural and substantive (see Everton & Hughes1987). Under the first heading there were two main complaints against the Inspector's handling of the inquiry. First, that he had refused to entertain the inclusion of areas outside the Council's preferred areas; and secondly, that he refused to hear arguments about land being wrongly excluded. Under the latter heading the essence of the objector's case was that the Council, in using the sieve map approach, had started from the basis that agriculture, landscape and amenity had to be protected and that what the industry would receive would be the residue. On the first hearing of the case the Court upheld the objections, with the judge arguing that some of the constraints used in the process were "subjective" (such as the locally designated AALs) and were therefore inappropriate. However, as the result of an appeal, heard in 1984, the Plan was reinstated.

Very little time elapsed after the reinstatement of the Plan before the planners were forced to begin the exercise again. About six months after Buckinghamshire County Council published its Minerals Plan the DoE published Circular 21/82 showing how demand in each region could be met over the 11-year period 1981–91. For the SE the most important feature of this Circular was that it anticipated that land-won sand and gravel would decline as a proportion of regional demand from 63% in 1977 to 54% in 1991 (quoted in Bucks County Council 1989b para. 30). However, although this proportion had declined in *absolute* terms land-won sand and gravel production was around 3 mt higher in 1987 than in 1977. Actual demand had exceeded the 1980 DoE forecasts which had laid the basis for the guidelines (Bucks County Council 1989b: para. 31). Consequently each RAWP was asked to prepare a commentary on how demand could be met up to 2005.

The South East RAWP (SERAWP) published its commentary in 1987. Although SERAWP considered that demand in the South East would continue to rise, it saw no increase in the levels of land-won aggregates coming from the region and stated that "a continuation of the 1985 level of production would seem to be a cautiously optimistic assumption" (quoted in Bucks County Council 1989b: para. 36). In March 1989 the DoE published revised guidelines in MPG 6 which endorsed SERAWPs conclusion and reiterated that the land-won total would be 32.5 mt (the 1985 level) (DoE 1989a: Annex A, para. 10). Bucks County Council had once again, therefore, to draw up a new plan stating how it would meet its share of the regional total. It was also incumbent upon the Council to maintain a landbank "sufficient for at least ten years extractions" (DoE 1989a: para. 34). In Buckinghamshire's case this meant a landbank containing 13 mt to last up to the end of 2001. In the new plan, therefore, a means had to be found to bring forward sufficient permissions to maintain the yearly quota *and* the landbank. To attain this "new consents totalling at least 18.08 mt would need to be granted in the plan period" (Bucks County Council 1990: para. 29). For this figure to be reached it was clear that the burden of mineral working could not continue to fall on south Buckinghamshire for it was recognized that "unless further releases are made [in mid and north Bucks] it can be expected that by the late 1990s the whole of the county's output of sand and gravel will come from south Buckinghamshire" (Bucks County Council 1990: para. 53). According to a Planning Officer who had worked on the Plan: "The question really comes down to whether we protect the rural environment or the semi-urban environment of South Bucks. There is a feeling in the south of the county that they've taken the brunt of minerals development. There is a political will to shift the balance, to find sites outside South Bucks".

The only way this "balance" could be achieved was by releasing more sites in the north, but as most of the sand and gravel deposits were in the Ouse Valley Area of Attractive Landscape (AAL) this meant the possibility of undermining one of the Authority's key conservation constraints. According to the above Planning Officer, taking this decision resulted in "much heart-searching" for the planners and Planning Committee members. However, during 1990 the Council lost two appeals, both the Inspectors prioritizing the "regional demands" argument, and one of the sites was in an AAL. It was clear that if they were not to lose more appeals, something planners generally abhor, they would have to address the shortfall in sites. The decision was therefore taken to open up the AALs and three sites were identified as "preferred areas"; (this was possible because there had always been a "needs" clause in the AAL constraint). However, having designated these sites the Council attempted to quickly "close the door" by arguing that "although the County Council is proposing limited exceptions within the AALs it does not intend to allow any further exceptions within the period to the end of 2001 on an *ad hoc* basis" (Bucks County Council 1990: para. 74).

Whether this policy succeeds in creating a consensus where previous efforts have failed remains to be seen. What is not in doubt is that the pressure on those reserves that are accessible will remain intense. As a minerals planning officer made clear to us, each time the planning authority goes through the process of identifying sites it "scrapes the bottom of the barrel". Finding sites in the centre and north of the county during the next round will be even more problematic and controversial. Far from creating some form of consensus, this highly visible form of minerals planning seems to highlight the potential for conflict.

A further twist in this convoluted planning process was given as a result of the public inquiry into the 1990 Minerals Plan. The Inspector decided that, after all, the Plan should not *quantify* a landbank until 2001 (the ten year supply) but only until 1996. Certain identified areas should therefore be deleted from the Plan. Significantly, all the deleted sites were in the AALs and northern Buckinghamshire. The Inspector apparently took no account of the need to shift mineral working from the south to the north of the county. However, in 1996 the Council will have to go through the whole process of identification once again in order to quantify a landbank until 2001. As a slightly exasperated minerals planning officer said: "It's a complete mission impossible; before you complete the cycle someone changes all the ground rules".

The minerals planning process in Buckinghamshire can be characterized as a struggle to find a political compromise between the need for aggregates and the demands of local residents for an undisturbed rural environment. This has driven the Minerals Planning Authority to adopt formal minerals planning procedures which involve a great deal of public consultation. The sheer scale of the pressure on the Planning Authority from local residents, action groups and conservation organizations means that the Authority must formalize its decision-making procedures in order to try and legitimize the trade-off between the regional demand for aggregates (as negotiated through the SERAWP) and local residents' demands for "peace and quiet" in the countryside. The problem with this approach, however, is that it makes the whole process more visible and therefore more susceptible to open contestation. For example, the County

Council received 1,200 objections to the draft version and 3,300 objections to the deposit version of its latest minerals plan.

Having established the overall planning context, we can now go on to explore how these competing pressures play themselves out in practice by following through a case study of one minerals extraction planning application. This shows how the various participants in the process represent their interests and how the planning authority seeks to arbitrate between them.

The minerals case study

The proposed development

This case begins in 1986 with the investigation by Steetley Quarry Products Ltd, a multinational quarrying firm based in the north of England, of a sand and gravel site in north Buckinghamshire at Chackmore, just outside Buckingham. An initial survey yielded positive results and a more detailed survey was carried out in 1988. Having established that there was a sizeable deposit suitable for extraction, Steetley then approached the landowners with the intention of securing their agreement to allow the company to extract minerals. The site included part of two farms which were both poor, one undeveloped and farmed very traditionally, the other a small family farm. Both sets of landowners entered into negotiations with Steetley over terms for extraction. The outcome of these negotiations was that Steetley took out a lease on the land and undertook to pay royalties on the minerals extracted (amounting, according to one informant, to £1 per tonne) and return the land to agricultural use at the end of the development (this clause had subsequently to be amended owing to political pressures – see below).

This marked the first of the relationships that Steetley would have to establish if they were to be successful. The landowners were incorporated into the scheme by the promised royalty payments. They were able to use their property rights as a resource to extract from Steetley a share of the minerals value. Although they were to lose the use of their land for approximately 14 years the financial gains would be considerable. This was for Steetley however only the precursor to the main struggle which was to bring the decision-making authorities and key political actors within a "pro-development" network.

The site was in an area of very attractive countryside, although it lay just outside an AAL so it was not under any designation. It was within an area of "search" and was formally "available" for mineral working. However, the site was close to Stowe Park which was leased by the National Trust. The grounds of Stowe Park were designated as a Natural History Site and the Stowe area had been designated as an AAL. The development site was also bounded by Stowe Avenue which formed an approach to the Park, and by roads to nearby villages.

The proposed area of development extended to 45.4 ha situated in undulating farmland which was primarily used for the cultivation of arable crops. According to the MAFF Land Classification System the site was designated as being a combination of grades 3a, 3b and 3c land, with the majority being in the lower grades. There was no outstanding ecological value in the majority of the site, although there was something unusual about part of one of the farms because one of the farmers had not "improved" the farm in line with modern production methods. The farm buildings had been designated Grade 2 by the Historical Society and part of one of the fields contained many species of meadow plants forming a relic of ancient grassland.

Under the original application it was proposed that site operations would begin as soon as possible after the granting of planning permission and would last for approximately 12 years. The anticipated output would be 250,000 tonnes per year. The volume of inert wastes required for infilling would be of the order of 650,000 cubic metres over an eight-year period at an average of 82,000 cubic metres per year. Infilling would take place concurrently with the extraction of sand and gravel. The final stages of restoration and landscaping were to be completed within 12–18 months from the cessation of tipping. The main objective of the restoration was to establish the soil profiles and horizons that existed prior to excavation. The number of lorry movements to and from the site would be 60 return journeys per day.

Company strategy and local action

In this section we will look at the strategies and activities of the main actors – the developer and local activists, and those operating the decision-making system, i.e. planners and their political members.

The developer   In framing their proposal for the planning system Steetley obviously attempted to stress the positive reasons for the granting of permission. Their main argument was based on the "need" for minerals. Their planning brief, for instance, laid great stress on the regional demand for aggregates: "Latest information collated from the South East Regional Aggregates Working Party states that current sand and gravel output in the County amounts to approximately 1.3 mt per annum and an existing landbank of consented resources estimated to be 8.3 mt. Clearly there is a shortfall of reserves to maintain the minimum, 10 year, government guidelines." Furthermore:

> The 1980 County Structure Plan states that the area of North Bucks extending towards Northamptonshire has been assessed as an area of potential for substantial growth; and in due course is expected to

develop as an important self-contained city region. As a consequence there are large government commitments for rapid growth in Milton Keynes. It is felt that the Policy to encourage imported borne sand and gravel into the area has not been effective and therefore a shortfall of available material is being experienced.

The company argued that this shortfall was affecting their ability to provide construction materials to the surrounding market area (Steetley owned ready-mix concrete plants in the area and wanted the sand and gravel to supply these) and was therefore putting jobs at risk.

Steetley were obviously keen to deflect some of the criticism which would come from local residents, most notably over the aesthetic impact of the development. In their proposal they argued, for instance, that: "Due regard has been taken of residential properties in the vicinity of the site and proposals include screening with tree planting to ensure that the amenity is properly protected. The processing plant . . . has been positioned as a result of an impact analysis to take advantage of existing hedgerows and topography to maximize the screening effects".

Having formulated their plans, however, the company then had to follow a political strategy to get them implemented. This meant negotiating the planning system and the "needs" argument was at the forefront of Steetley's presentation of their case to the planning authorities. Having put their case to the authorities, however, Steetley had then to attempt to sell it to the local community and this was a far more difficult task.

According to a local community activist who opposed the development, Steetley's approach was straightforward and had been well tried elsewhere; they put in their initial proposals, without much research into the likely local problems, then waited for the objections to come in. They then amended their proposal to take account of those objections. This strategy had two main advantages for the company; first, it saved them extensive research in the area and secondly, it allowed them to present themselves as being very receptive to the demands of local communities.

After lodging their initial proposals with the planning authorities the company was obliged under Section 26 of the 1972 Town and Country Planning Act (the "bad neighbour" provision) to advertise their intentions locally, and they did so two months later by placing a notice in the local press. This left only six weeks before the Planning Committee were due to consider the proposal. Nevertheless, as expected, after the company's intentions became known local opposition began to form. The company responded by setting up a caravan in Chackmore in an effort to try and reassure local people and provide information on the company's "good" work practices. This was normal practice for the company when attempting to get a foothold in a place where they had no history of activity. By all accounts they received a fairly hostile response.

Steetley then called a public meeting in Buckingham to discuss the proposals. According to a local opponent who attended the meeting it was "very emotional and badly run". For instance, he claimed that Steetley hired "bouncers" who became involved in an ugly fracas. After Steetley had presented their case they opened up the meeting to the floor. There were quite a few condemnations of the proposed development. Then a young man who had been partially crippled by a gravel lorry stood up and tried to speak. At this point the chairman decided he would hear no more comments from the floor. The young man insisted on his right to speak and said he would do so from the platform. As he attempted to make his way to the platform the "bouncers" intervened and tried to return him to his seat whereupon the "bouncers" were themselves set upon by some of the young man's friends. A scuffle ensued, at which point the chairman realized things were getting out of control, changed his mind and said that the young man could say his piece after all. Having reached the platform the young man then proceeded to show the meeting his crippled leg, saying that this was the result of gravel lorries on the roads. All-in-all the meeting does not appear to have done much for the company's image in the locality.

All this took place in the summer of 1989 and during this period the number of objections grew. The most serious of these concerned the ability of the local roads to withstand heavy lorry movements and the effects of the gravel pit on Stowe lake (it was argued by the action group that the extraction of minerals would lead to a lowering of the water table and would therefore affect the ornamental lakes, the centrepiece of Stowe Park – we shall explore this in more detail below). In the face of these objections Steetley withdrew their original application for amendment.

On March 26 1990 Steetley submitted their amended proposals to the planning departments. The revisions were as follows: (a) to deal with the threat to the water table the excavations would not be as deep as originally planned. The excavation would take place over a 14 year period with the annual anticipated output reduced to 180,000 tonnes; (b) to counter the objection that the roads were unsuitable, traffic was to be re-routed; (c) the number of lorry movements was reduced to 86 (43 return journeys) from the 120 originally proposed; (d) the reduction in traffic movements was to be achieved in part because no waste would be imported to infill the site. The land could not therefore be restored to its original level and it was proposed to create two lakes and a conservation area; and (e) the ancient grassland would be transported to a new location to the northeast of one of the farms. This revised application was scheduled to be considered by Buckinghamshire County Council planning subcommittee in April but any decision was deferred by the committee until June.

In the meantime Buckingham Town Council met Steetley's management to discuss the new proposals. Steetley argued that the new road routes, which entailed all the lorries going through Buckingham, were under-used according to Department of Transport estimates. The Buckingham Advertiser (20 April 1990) reported that this information met with little sympathy from the Town Council. The latter argued that the road system in the area, and specifically in the town itself, simply could not cope with the extra traffic which the development entailed. Steetley's representative responded by claiming that the company had taken local opinion very much into account – hence the amendments to the original plan. At this meeting the Town Council decided, with Steetley's agreement, to organize a public meeting and invite people from the other parishes likely to be affected.

This public meeting took place at the beginning of May. Once again Steetley were asked to present their case. Their first speaker was from the company's corporate communications department. He repeated the argument that Buckinghamshire was falling behind its annual target of 1.3 mt of sand and gravel and that too little was coming from the north of the county. He said the Chackmore site was chosen because of its rich deposits of sand and gravel, because it was not Grade 1 agricultural land, and because it was not in a designated Area of Attractive Landscape or Site of Special Scientific Interest. He added that the company always listened to constructive suggestions and had scaled down its proposed operations at the site to reduce traffic movements, thereby avoiding lorries going through a nearby village. The next spokesman for the company was a traffic consultant. He said a survey conducted on one day in the previous October showed that Steetley's lorries at 83 traffic movements a day would mean a traffic increase on local roads of between one and two per cent. However, questioned by the audience, he admitted that the increase in the number of lorries of over 7.5 tonnes using these roads would be 57%. The consultant added that using Department of Transport guidelines both roads had spare capacity. He further added: "Certainly there will be a marginal increase in congestion under these proposals, but I think that will be insignificant; you are talking about an increase in traffic that will hardly make any difference" (*Buckingham Advertiser*, 4 May 1990). In response to this a Buckingham District Councillor voiced her concern about lorries using one particular road as it could barely take two cars passing. Steetley's representative said the company had talked to Bucks County Council about widening the road and possibly doing some work at the access point to the A413, although what work was involved was not specified. A Town and County Councillor then asked: "Has it occurred to Steetley that we do not want this road widened". According to the *Buckingham Advertiser* this was met by general applause.

This meeting has been documented at length here (the details come from the *Buckingham Advertiser*, 4 May 1990) as it illustrates the kinds of arguments Steetley were using in public to justify the development and the difficulties they faced as local antagonism grew. These difficulties were fully recognized by the company. As a manager involved in this public consultation exercise put it: "At the end of the day you are in a no-win situation because nobody wants you. But at least you can put your case forward. I think at Chackmore that we put our case very well – we were told that even by the 'antis': 'we think you are all great guys and you are doing an excellent job but we don't want you' That's peoples' philosophy and its very understandable". In this situation all Steetley could do was try to play down the effect the development would have on local communities. They went to great efforts in devising their plans to ensure that the gravel working would not be visible either from local properties or from the important approach to Stowe Park. Clearly this "minimal impact" strategy had its shortcomings and could not really stifle the rising tide of opposition. But the company had little alternative, as the above-mentioned manager explained: "If you can't point to anything tangible except perhaps road improvements, or that it's derelict land and you've got to take it up and make it better than it was, then you really are stuck with the 'need' aspect and the fact that there is a need in the community. Its very hard to sell that argument I'm afraid".

The latter argument met with little sympathy in the local community. The discourse around "need" was in effect for the planners for it was they who had to balance the competing arguments. This left Steetley exposed in their public consultations. However, they attempted to present themselves as aware and responsive to local concerns by again amending their proposals. Once again this was stressed by the manager: "When we went through this public information exercise people came up with one or two valid queries. On the basis of peoples' concerns we went away and did the background work and we've answered every one of those issues". All Steetley could hope to achieve during the public consultation period was some minimising of the opposition. It was the planning authorities, the decision-makers, who were the real target of Steetley's campaign. Here they were much more influential, as we shall see below.

The local action group The first time the local residents became aware of the proposed gravel pit was when they read the statutory notice in the local paper. Their initial reaction, according to a local campaigner, was anger because no-one had told them about it despite the fact that the proposal had been lodged with the planning office two months previously. This local resident and a friend of his who also lived in Chackmore went to the planning department to read the proposals. Having done so they then began to phone round the affected villages and discovered (or created) a great deal of concern. They organized two people in each village to take the matter up with the parish councils. However, the response from the parish councils was mixed, with some opposing the development, others supporting it, and some expressing no real concern. Owing to the lack of leadership from this quarter the local campaigner and his colleagues in the other villages decided to do something. They formed a committee which included two people from each of the communities concerned (these were all, incidentally, "professional" people) and decided to prepare a report to present their views to the planning authorities. Each village was allocated a task: for instance, one wrote a history of the area, while another looked at the traffic problems, using the advice of a policeman, and a third explored other problems associated with the development. The group decided to adopt the name Chackmore Against Gravel Extraction (CAGE).

There seemed to be one campaigner in Chackmore who took the lead in the campaign against the proposals. He was retired, after having worked as a marketing professional, and was thus able to put a lot of time and energy into the campaign. He also worked quite closely with the bursar of Stowe School (a top-notch public school). Both these campaigners were aware that a strong case against the proposals could be mounted if they found evidence of any possible effects on the ornamental gardens in the school grounds, designed by Capability Brown. These gardens had recently been leased by the National Trust who intended to spend around £2 million on their restoration over a ten-year period. This activist therefore called in a hydrologist friend to identify any likely threat to the lakes which form the centrepiece of the gardens. The hydrologist pointed to the effect of the proposed gravel working on the water table and postulated that this could reduce the supply to the ornamental lakes. When the potential hydrological problems with the site became known to the campaigner he told the bursar but advised him to keep this information quiet until just prior to the planning committee's consideration of the proposal (believing that it was Steetley's strategy to neutralize objections by amending their proposals). The bursar, however, did not take this advice and lodged an objection immediately, thereby alerting the National Trust to the potential threat to the lakes.

The action group drew up a report listing their objections to the development. However, they did not release the document or lodge it with the planning department. Meanwhile, Steetley withdrew their initial application. The company then hired an independent consultant to look at the water problems. On the basis of the consultant's report, and the objections that had been put forward, Steetley amended their proposals.

CAGE likewise revised their report in the light of Steetley's new proposals in March 1990. They argued that the company could not deliver on the promises made about noise and pollution control, having seen Steetley's other quarries. In the light of their own traffic survey CAGE regarded the traffic problems as insurmountable without extensive road improvements which, they claimed, were opposed by local residents. Furthermore, they accused Steetley of "dishonesty" over the water issue. According to the CAGE report, Steetley's amendments did not adequately address the problem. They quoted the consultant's report as saying that even with the new level of working "the possibility of affecting the school's water supply springs is much lower but cannot be completely ruled out". They also expressed concern at the position of one of the lakes on the extraction site which they claimed was only 100m from Chackmore's nearest house and therefore posed a threat to village children.

In keeping with their general strategy, CAGE withheld this report until the last moment (14 days before the planning committee were due to consider the proposal at the beginning of June). Copies were then sent to the County Council planning department and the highways department, the local MP, George Walden (Con), and the ten Conservative members of the County Council's planning subcommittee. The aim was to target the most influential people in the planning process. The leading CAGE activist claimed that the report did lead to the campaign group being taken more seriously and cited as evidence the fact that the Assistant Chief Planning Officer, who had originally refused to see him, changed his mind after receiving the report. At this meeting he said the Planning Officer listened to him very politely and took note of his objections. The fact that this meeting took place at all was seen by CAGE as evidence that their report was having an impact. It is interesting to note, however, that following the presentation of the CAGE report, the Chair of the Planning Committee wrote to the activist asking him to verify that he was actually speaking officially on behalf of the local residents. Questions were raised about his representativeness; for whom was he actually speaking? The campaigner responded by going to the local parish councils to get letters of support for the work of CAGE. There is no doubt that this activist was the prime mover behind CAGE and the authorities were questioning his right to speak on behalf of those he claimed to represent. By mustering the support of the parish councils he was able to defend CAGE against the accusation that it was a "one-person show".

CAGE also linked up with other groups fighting quarries in North Oxfordshire and Northamptonshire. The three groups met their constituency MPs at the House of Commons. At this meeting three sets of objections to the conduct of minerals planning were discussed: (a) the way the planning system attempts to deal with proposed quarries in rural areas was inadequate, particularly where roads were unsuitable; (b) there was a lack of local democracy; and (c) nobody was doing anything about the really serious environmental issue which was the limited amount of gravel in the country. On this latter point the action groups argued that the real issue was: can we continue to dig up the countryside in pursuit of diminishing amounts of gravel? The MPs listened to the campaign group's case and promised to take up the issues with the relevant Minister.

175

The last of the three objections was the key issue for CAGE. They believed that it was only by addressing the whole approach to the extraction of minerals that the pressure on individual mineral sites could be decreased. CAGE believed they had an answer to this problem. Standing outside the coal mines is 1,000 mt of rock – this could be recycled as a minerals substitute. Standing nearby the power stations is 30 mt of fly-ash and there is a system currently available which turns fly-ash into high-grade aggregate. These two sources, according to CAGE, could meet 50% of the UK's aggregate requirement. The objective was to reduce the amount of quarrying that needs to be undertaken in the countryside, a necessary part of the fight against specific developments. Whether these proposals are at all feasible (a Steetley manager we interviewed said they were simply not economically viable) is unclear. They can be seen, however, as an attempt to shift the debate away from a simple counterpoint between local opposition and national "needs". The group hoped that as a result of this meeting the MPs would tackle the DoE on these wider considerations. CAGE also hoped that George Walden, MP for Buckingham, having received a copy of their report, would directly lobby Chris Patten, then Environment Minister, over Chackmore. The group was later informed that the MP had phoned the Minister to discuss these issues.

The strategy of the local action group was quite clear: on the one hand to stir up as much local opposition as possible while on the other trying to draw "national actors" into an "anti-development" network. Political actors with local bases, such as the MPs, were obvious targets, as was the National Trust with its interest in Stowe Park. Less obvious, but perhaps just as effective was an "old boy network" emanating from Stowe school. Certain members of this network, we were informed, hold seats in the House of Lords and similar institutions. Another "national" figure, Sir Ralph Verney, Chairman of the Advisory Committee on Aggregates, also became involved when he wrote a letter to the local press condemning the proposal as unnecessary. The issue had truly reached "national" proportions when an item appeared on Channel 4 News on 30 May 1990 showing the "threat" to "perhaps the finest example of English landscape gardening in the country". According to the National Trust spokesperson interviewed on the programme: "Stowe is the cradle of landscape gardening"; putting a gravel pit next to it would be like "splashing the Mona Lisa". The main concern expressed on the programme was the effect of the quarry on the lakes. The gardens had been designed by Capability Brown around the lakes to reflect the landscapes of 18th century watercolourists. Any threat to the lakes would effectively threaten the Park as a whole. Given the National Trust's commitment to the gardens, the development proposal was described as ill timed. Shortly after this programme the Royal Fine Arts Commission came out publicly against the proposal. Opposition among "national" actors was mounting.

It is perhaps therefore not surprising that in mid-July 1990 the DOE called the case in, effectively taking the proposal out of the hands of the County Council. The matter would either go to a public inquiry or would be turned down by the Minister. However, before this decision was made Steetley withdrew, claiming that a public inquiry would be too costly (*Buckingham Advertiser* 20 July 1990). The scale of the opposition and the adverse publicity generated by the development seemed, in the end, to have proved too much for the company.

Local "anti-development" actors undoubtedly generated a lot of opposition to the proposed development among the resident population. This may not however have been sufficient to kill the proposal. Had local politicians taken the decision the outcome could have gone either way (see next section). What proved crucial was the incorporation of national actors into the process. This incorporation hinged on one simple question: "would the proposed development damage the ornamental gardens at Stowe?" It was Steetley's inability to satisfactorily answer this which allowed an "anti-development" network of actors, extending from the "local" to the "national", to come into existence. In the deliberations of the National Trust, the Royal Fine Arts Commission and Stowe School's "old boy network" it was not the effect of the quarry on local life or the roads which was the issue but the consequences for a "national treasure". One of the local activists interviewed felt it had not been their campaign that had led to the outcome but the enrolment of the National Trust and Stowe "old boys" into the anti-development network. These two key actors had been able to apply pressure in just the right places (according to one respondent, well placed in the local planning authority, a "huge" amount of political pressure came from the National Trust; the "calm confidence with which the trust were talking was unbelievable"). It seems likely that Steetley had foreseen all the objections except the effect on the lakes. This dimension of the development, above all others, brought the National Trust into play. Steetley simply could not match the Trust at the national level. All their arguments about "need" were, in the end, overwhelmed by the perceived threat to Stowe Park.

The planning process While the ultimate decision on the development was taken out of the planners' hands it is interesting to speculate whether they would have been enrolled into the pro- or anti-development networks. No definitive answer can of course be given, but if we examine the planning procedures and the kinds of assessments being made by participants in the planning process we may get some indication.

This proposal undoubtedly put the planners under considerable pressure. The planning officers who dealt with the application at the County and District levels said they had never had to deal with such a complex set of issues. The planning departments were flooded with objections to the proposal and they were well aware of the strength of local feeling. These local issues had somehow to be balanced against the regional demand for minerals. In what follows we will sketch out how the planners attempted to achieve this "balancing act".

Initially Steetley submitted the application to the planning departments at both the District and the County levels. Under "normal" procedure, the decision would have been made by the Mineral Planning Authority (the County Council). The rôle of the District was merely to offer an opinion on whether permission should be granted. After Steetley publicized the proposal objections came into both offices. At the District Office a planning officer was assigned to the case whose rôle was to go through the proposal and pick out the key features, then to go through the objections and categorize them. This entailed the "translation" of a whole range of objections and comments, sent to the planning office in the form of letters, reports etc., into a set of manageable categories which allowed this material to be summarized for the Planning Committee members. The text also included a summary of the official objections of a number of local bodies. Almost all the local parish councils objected (once they became aware of local opposition). Their feelings were perhaps expressed most cogently by one which said: "[We] view with complete horror the proposal to open a gravel and sand extraction pit near Chackmore. On environmental, ecological and social grounds, [we] affirm [our] intention to strongly oppose any such planning application within this area" (statement issued on 22 May 1989). The local Town Council voiced concern on the effect increased heavy traffic would have on buildings in the town; the Landmark Trust objected, as they were negotiating with both Stowe School and the National Trust to take a long lease on the Corinthian Arch (apparently used as a model for the Arc de Triomphe) which stands at the entrance to the park; the National Trust voiced their concern about the effect of the pit on Stowe lake and complained that the development would fundamentally spoil the approach to Stowe Park; and the Berkshire, Buckinghamshire and Oxfordshire Naturalist's Trust (BBONT) stated that the development would damage wildlife.

Given the great weight of opinion against the development it is not surprising that the District Council, which had no need to take the issues associated with demand into consideration, should decide that planning permission should not be granted. This recommendation then went forward to the County where it would be taken into consideration by the County Councillors on the Planning Committee.

Such was the extent of the District's involvement; attention now focused on the County Council. The procedure here was much the same as at the District Council: a planning officer was assigned to the application and wrote a report weighing up the benefits against the objections, concluding with a recommendation for the Planning Committee. The considerations to be addressed were, however, more complex, for the "needs" arguments being pushed by Steetley, which we outlined above, must be taken into account. Furthermore, this matter was being considered following the loss of two appeals and against the background of the preparation of the County's latest Minerals Developments Plan which had opened up some of the AALs in an attempt to push minerals extraction into the north of the county. In fact the minerals plan was published in the midst of consideration of the Chackmore proposal and its recommendations were seen by Steetley as perverse. As a Steetley manager put it:

We've got the Stowe site which is not within an Area of Attractive Landscape, which 'does not have any prohibitions. They've now come out with a new minerals plan which has one or two sites in an AAL, which Chackmore isn't. Its just illogical. We find a site with a deposit, we approve a deposit which is a hidden deposit, you can't see the site, we find it and then they go and pick three or four proliferations of sites with half the reserve and instead of having one site without any restrictions they go and plant them in an AAL. Now that to me does not stand in line with what the policies are in place for.

This is the kind of argument that the planning officers were well aware would be levelled at them at a public inquiry. Furthermore, such considerations had more salience at a time when the county was falling behind in its minerals requirements. In presenting the report to the planning committee therefore, the planning officers had somehow to weigh up the list of objections to the development against the regional requirements for aggregates and all these issues would be weighed against one another in the committee by the lay members who would make the decision. While it is quite possible that, swayed by the vociferous campaign being waged in the District, they would have come out against the proposals, certain comments amongst planning staff indicated that the proposal was, from their perspective, viable. Planning officers, in interview, expressed surprise that Steetley had withdrawn and felt that their chances of success at a public inquiry were by no means negligible. They believed that the case against the development, particularly the threat to Stowe gardens, was "not proven". Further evidence of some sympathy for Steetley in the upper echelons of the department was provided by a local "anti-development" campaigner who, in a meeting with a senior planning officer, was told that the campaign group was perhaps being too hard on the company, who had gone out of their way to meet the demands of the local population. Lastly, we were subsequently unofficially informed that the Planning Officer's recommendation to the Planning Committee would have been to approve the scheme.

Such anecdotes of course prove nothing, but they do indicate the strength of the "needs" argument in a case which was extremely unpopular in the locality. The attempt to "balance" sets of local considerations against assessments of strategic demands is fraught with difficulty. In this case the regional demand for minerals may well have overwhelmed local "quality of life" issues. However, local actors were able to marshall "national" concerns for a park of unique importance to the nation, thus recruiting "national" actors (crucially the National Trust) to the anti-development cause. Eventually the Secretary of State for the Environment was drawn in on terms set by the anti-development network (concerns about the lakes), leading ultimately to the proposals' demise.

Back to the context: subsequent debates in minerals planning

This case study highlights the rôle of regional planning in the decision processes surrounding a particular development. The local opposition to the development proposal began to develop a critique of the forecasts which underpin the regional and county aggregates targets. These forecasts returned to the headlines in May 1991 when the government's latest assessment of the need for sand, gravel and crushed rock showed that minerals extraction could increase by 66% up to the year 2011. According to these projections, annual aggregate production in the UK will be 500mt by 2011, as opposed to the 300mt being won each year at present. These calculations were attacked by the CPRE as "flawed" and lacking in any consideration of the environmental impact of an aggregate winning programme on this scale (*Planning* 916, 3 May 1991). A report commissioned for the CPRE (Adams 1991) criticized the "predict and provide" philosophy of the minerals planning system as outdated and environmentally insensitive. It pointed out that the 1976 forecasts predicted a level of aggregate demand 1980–85 which was 50% above the actual rate of output. The 1988 forecast was equally erroneous – but in the opposite direction – within three years of its publication:

> Sometimes aggregates demand runs ahead of construction investment, sometimes it lags behind. Sometimes construction investment goes down when GDP goes up. It is not surprising, therefore, that by 1989 the forecasts published in 1988 were out by about 50%. What is surprising, given the forecasters failure to identify a relationship between economic activity and aggregates demand, is their decision to rely on a model which has just such a relationship at its heart. (Adams 1991: 11)

This is of particular concern because: "Local planning authorities cannot challenge the national forecasts, only the amount of minerals they are expected to produce towards meeting total forecast demand. Anyone objecting to the granting of planning permission for aggregates extraction must confine their objection to the particular site proposed. They may not challenge the total demand forecasts, and are in effect restricted to arguing 'not in my back yard'" (6).

Following publication of the 1991 forecasts the Government went to some lengths to stress that "the forecasts do not represent firm Government targets". Junior Environment Minister Tim Yeo said: "While the forecasts provide general background information about trends in demand, their use in relation to the consideration of a particular application or development plan is necessarily limited until the revised guidelines are published" (*Planning* 917, 10 May 1991).

The Chackmore case coincided with this "national" debate and, after consultations with CAGE and CPRE, George Walden, MP for Buckingham, raised the issue in the House of Commons on May 9 1991. He claimed that:

"In huge swathes of north Buckinghamshire people are in danger of having their lives blighted and the countryside is in danger of being irretrievably scarred by the operation of gravel diggers . . . . The village of Chackmore in my constituency is a charming quiet village quite close to Buckingham in the environs of the historic grounds of Stowe. It is incredible to believe that an application was put in for gravel extraction within hundreds of yards of that village. It was fought, but it was only after a campaign in the national press that the application was withdrawn. Not everyone has the pulling power, in terms of media interest of Stowe school, or the support of the National Trust" (*Hansard*, 9 May 1991: 896).

Furthermore: "We seem to be caught in a vice on this matter. We seem to be in a position where we have to despoil the countryside to improve the built environment; we have to blight the lives of rural residents to improve urban areas, housing and roads" (897). The MP was however keen to emphasize that he was not just speaking for his rural constituents: "We seem to be in a position where people who live in urban areas and who might want to seek a little tranquillity in the countryside at the weekend will find dust, noise and gaping holes in the ground in the villages that I have mentioned. It seems to me to be in everyone's interests to re-examine the whole question of extraction of aggregates and the implications for the environment" (897). Turning to the minerals planning system Walden argued that it

. . . all sounds neat and immaculate as a piece of planning machinery, but, in practice, it has become completely divorced from the results on the ground. It is all very well to have RAWP and SERAWP and means of co-ordinating them, but something has gone alarmingly wrong with the whole bureaucratic structure when it ends up in a serious proposal to place a gravel extraction pit next to a school, a convent or Stowe school. (898)

In reply the Parliamentary Under-Secretary of State for the Environment, Tony Baldry, repeated the Government's reconsideration of the use of the forecasts: "I must stress that the forecasts do not represent Government plans, and are not targets for production which the minerals industry must meet. It is important, therefore, that they should be considered as defining a problem, rather than providing a solution. We regard the environmental implications of the forecasts very seriously, and intend to conduct a full and open debate on the planning and environmental issues they raise" (903).

The initial response from the minerals industry was to make a plea for calm. The President of the Sand and Gravel Association (SAGA) said: "Our hope is that the Department of the Environment and all those concerned will show the clarity of thought, sense of purpose and vision to work together to see that an adequate and regular supply of aggregates is achieved at the best balance of environmental social costs" (*Planning* 920, 31 May 1991). The trade association British Aggregate Construction Materials Industries (BACMI) came out four-square behind the forecasts, claiming that "they are the most thoroughly prepared yet and . . . they will be a necessary tool for future planning". BACMI rejected the accusation that the forecasts lead demand: "An over-high forecast cannot result in one more tonne of aggregate being sold, nor one unnecessary quarry being opened . . . . Quarrying can only respond to the demand for its products from the public through the construction industry" (*Planning* 937, 27 September 1991).

At the time of writing this row looks set to continue. The question of catering for growth within the (SE) region seems to have focused attention on both the search for alternatives, as proposed by CAGE, (the CAGE proposals were also mentioned by Walden in the Commons debate) and the need for efficiency (as in the latest CPRE proposals for a "Minerals Efficiency Office" – see *Planning* 937, 27 September 1991). What seems clear is that the opposition to land-won aggregates from within the region is strong, at both local and national levels.

A short-term solution to the imbalance between supply and demand in the SE in the form of other sources of supply is currently being considered by the Government (MPG 6 forecast a rise in imports to the region of 100% between 1985 and 2006 DoE 1991). The Parliamentary Undersecretary of State for the Environment mentioned a DoE research project which is "examining the possibilities offered by large-scale coastal quarries, such as that at Glensanda in Scotland. It will examine the potential areas of supply, which include Scotland, and indeed Ireland, Norway and the

Iberian Peninsula. The consultants will also investigate the environmental and economic consequences of providing aggregates to the south east from such sources" (*Hansard*, May 9 1991: 904). This raises the spectre of "super-quarries" in Scotland providing aggregates for construction industries in the SE. The aggregates industry is already talking of seven coastal quarries. The Western Isles is considering a proposal from Redland Aggregates to extract 600 mt of rock within a "national scenic area". The deputy manager of Redlands claims coastal super-quarries are "the only realistic solution" to the problems posed by environmental constraints on English quarries (*The Guardian* 7 June 1991). The extraction of aggregates becomes someone else's "local" problem. This time, however, there may not be well organized action groups in Conservative constituencies to argue the case against development. Furthermore, arguments in favour of creating local employment may carry more weight in Scotland than in places such as rural Buckinghamshire.

Conclusion

In the preceding case study we have described the process of development as a contest between two sets of actors: one pushing a particular set of production-oriented values while the other countered with what can loosely be described as a consumer aspirations concerned with "quality of life" issues and concern over the nation's heritage. The former consisted most prominently of the landowners who, through their ownership of private property rights, were able to negotiate for themselves a substantial financial gain. This could only be realized, however, through the development company. The ownership of mineral rights is of little value unless someone is interested in extracting the minerals. For the development company the acquisition of such rights is only the first stage in the long process of negotiation in which they must become involved in order to actually dig the material out of the ground, sell it and make a profit.

The quarrying company Steetley must be seen as the prime mover in this process. They had the resources, the motivation and the required knowledge of the rules to put the development process in train. It is important however to put their action in context. As we noted earlier, the SE generates particular demands for aggregates. These demands are structured by the regional minerals planning framework in which the minerals industry is strongly represented. This framework can be seen as an attempt by the developers to structure the rules in a way which meets their perceived interests. The implications of this set of rules for particular site-specific planning decisions are also referred to above. In their representations to the planning authorities, Steetley were able to draw upon an existing set of arguments surrounding "needs", which gained their legitimacy from the whole complex institutional structure which made the "needs" assessments. Furthermore they were doing this at a time when such arguments were at the forefront of the planning authority's considerations. Steetley then abided fairly strictly by the planning rules, for these offered the best chance of achieving the desired outcome. Although they went through the motions of a public dialogue they concentrated their resources on presenting their case where the decision would be taken – the Planning Committee of the County Council.

The local action groups, formed in response to Steetley's proposal, soon learned, however, that these planning rules could well be working against their perceived interests. It is noticeable that by the end of the process they were attempting to raise questions about the operation of the rules with those who had some part in formulating them. In order to circumvent the operation of a set of rules which seemed to offer an uncertain outcome the local actors attempted to draw into their anti-development network actors who were uniquely placed to go straight to another "power point": the Minister. Once again we must place the actions of this group in context. The social changes outlined earlier have allowed the emergence in the Buckinghamshire countryside of ex-urban retired or commuting professionals. These people generally have the cultural and economic resources to represent themselves effectively within political processes. They can often draw upon levels of expertise in fields which are not part of the "traditional" rural community. For instance, it was noticeable that a crucial intervention was made by a member of the action group when he called in a hydrologist friend who identified the "threat" to Stowe gardens. This, as we have argued above, shifted the balance of forces as it mobilized a set of strategically placed actors against the proposal, eventually resulting in its defeat.

In some respects the story of the Chackmore proposal may seem unique but we believe it serves to illustrate the process whereby competing sets of actors attempt to impose their representations upon the development process. Although this is necessarily a "local" struggle, as we have seen, strategic arguments and concern with the "national heritage" sometimes take their place alongside "local" concerns such as the state of the roads and the effect on local communities. It seems to us that how and why these representations came into existence, and took the form that they did, can only be understood by examining the process of negotiation. This will of necessity always take a unique form but by situating the specific case within the wider context, and by making reference to the rules and patterns of representation, we gain a much deeper understanding of how that context itself comes into being. The case study highlights further how the context changes; it is not static but constituted by the multitude of actions which reproduce it. This reproduction does not however lead to its continual replication, for it comes about through the strategies of, and contests between, actors. These actors may be seeking to reproduce particular organizational patterns (part of what we see as the "context") but they may also be seeking to challenge such patterns. Sufficient challengers, commanding sufficient resources, can change the context irrevocably, setting the stage for the next "round" of contestation. The Chackmore case seems to have played its part in shifting the organizational structure surrounding minerals extraction. While the case study demonstrates how "national" considerations enter into "local" development processes, it also indicates how "local" issues enter into "national" political processes. It was used by the local MP to highlight how the minerals planning system "has lost all touch with the lives of ordinary people" (*Hansard* 9 May 1991: 897). There is no doubt that the political structure surrounding minerals extraction will attempt to conduct business as usual as far as possible. But in the rural areas of Buckinghamshire it is clearly becoming difficult to find enough sites to meet the construction needs that have traditionally been so high in the SE. And it is not just in Buckinghamshire that such constraints exist, they are widespread across much of southern England where there are many people capable of mounting effective campaigns against unwelcome development. This may result in coastal super-quarries in Scotland exporting minerals to the SE. Thus, constraint on development in one place may result in potential overdevelopment in another.

CHAPTER 8

Land to waste

Introduction

Unwelcome development in the countryside is not just confined to minerals. Another sector that awakens a great deal of controversy is waste dumping. While the extraction of minerals is generally regarded as being the kind of "dirty" and "noisy" activity that many people now living in the countryside find wholly undesirable, the view of landfill or waste dumping is probably even more negative. Minerals at least have the virtue of being "useful"; landfill deals with the discarded and the "useless". Moreover, this "useless" material is often toxic and dirty. A landfill site is definitely not the sort of thing that most people want "in their backyard". In our social survey only 4.2% of the total sample wanted to see more waste dumping in the countryside while 62.1% wanted less (for the villagers the figures were 4.7% and 65.4% respectively). The nub of the problem is, therefore, where are landfill sites to be situated?

In the previous chapter we showed how "national" demands for rural resources (minerals) were balanced against "local" concerns in the development process. In cases of minerals development, action groups find their "local" arguments being weighed against regional and national considerations associated with the "demand" for construction materials. What seemed to swing the Chackmore case against development was the participation of the National Trust who were attempting to protect a "nationally treasured" landscape feature. This allowed local middle-class residents to enter into a provisional alliance with a "powerful" national actor in order to prevent any development taking place. The alliance was crucial in determining the outcome, for it allowed "national" conservation considerations to be mobilized against "national" calculations of aggregates demand. While waste disposal does not bring to bear "national" considerations it does raise questions of "regional" significance i.e. where does all the waste go? Again, the regulatory authorities (the county council planning departments) have the task of balancing local opposition to development against more strategic concerns. However, the planning machinery for waste disposal is much less developed than for minerals. This, as we shall see, makes the task of the pro-development actors more complex for there are fewer strategic, extra-local arrangements to persuade the local authority to ignore the representations of the residents' action groups.

The planning context for waste

Traditionally landfill has taken place in existing holes in the ground: disused quarries, minerals workings or derelict land. The disposal of wastes was a means of "reclaiming" such land for a renewed, often agricultural, use. Past practice also linked minerals development permissions to the reclamation of the land through infilling. Nevertheless, there was still some concern at the environmental and health implications of this practice. The perception that many sites were not well kept gave rise to the 1974 Control of Pollution Act which created the Waste Disposal Authorities. The county councils were given licensing powers over the tipping operations of private companies, in addition to the responsibility for the disposal of refuse collected by the district councils. The Act also required the councils to draw up Waste Disposal Plans. Buckinghamshire County Council, although stating its intention to draw up such a plan in 1980 (Bucks County Council 1980), had by 1990 not done so. According to a planning officer, this was because all the proposals submitted to the authority were to fill disused pits and quarries and so were covered by the Council's licensing and aftercare statements. The establishment of Waste Disposal Authorities undoubtedly gave an impetus towards better controlled and managed sites but this was to some extent superseded by renewed fears over landfill sites: "The publicity over leachate production and more recently over the more insidious dangers of migrating landfill gas has added to the poor public perception of landfill to the extent that it is often viewed as dangerous, low technology, environmentally crude and aesthetically unacceptable" (Morrey 1990: 6).

The persistence of such attitudes to waste disposal have made it a highly unpopular activity in many countryside areas (the previous chapter indicated some of the current pressures to separate the "digging" and "filling" of holes). The long-term effect of this is likely to be increasing difficulty in finding new waste disposal sites. Although alternatives to landfill, such as incineration, are available for many types of waste, landfill is still regarded as the most economically viable. A 1987 SERPLAN report highlighted a growing problem when it said: "While there is no immediate crisis, the longer term problems of accommodating London's waste, together with that arising in the rest of the region, will increase as the suitable void spaces closest to London are used up; and from the mid-1990s it will increasingly be necessary to transport waste further, and at therefore greater cost, to disposal sites" (SERPLAN 1987: 2). This report also put forward the view that "the industry and local authorities should proceed

on the basis that disposal to landfill will be the best practicable environmental option for the majority of unavoidable waste". This type of waste
disposal has two components: filling existing holes and land raising. SER-
PLAN considered that

> priority should normally be attached to landfilling void spaces, and
> that land raising should only be considered if there is no reasonably
> practicable alternative means of disposal. Sites should not be for
> land raising unless it can be clearly demonstrated that: (a) the oper-
> ation would be environmentally acceptable; (b) the landform to be
> created would be visually acceptable; and (c) the ultimate end-use
> is acceptable and would be compatible with adjoining land uses.(5)

Such conditions are of a general nature and local interpretations of
them are likely to vary. We will examine a proposal of the "land raising"
type in Aylesbury Vale and the reaction and influence of the local community. This provides us with another empirical example of how local
actors organize themselves into action groups to prevent development
from taking place. Given the increasing difficulty of finding suitable
waste disposal sites in the region the conflict surrounding this proposal
would seem to be by no means atypical. Moreover, the increasing problems associated with finding holes to fill with waste implies that disposing of waste by land raising may become more widely used. However, as
this case shows, such proposals are likely to be fiercely resisted at the
local level.

Waste: the case study

In February 1990 Hales Waste Control company put forward an application to develop a waste tip at Bierton, a village just outside Aylesbury on
the A418 Aylesbury to Leighton Buzzard road. The proposed development site was on farmland just outside the village. The land was owned
by an insurance company and was farmed by a tenant farmer who was
also a District Councillor. The site comprised six fields (35 ha) of grassland adjacent to an Area of Attractive Landscape.

Hales, an independent company within the Ready Mix Concrete
group, is one of the largest waste control firms, working mainly in the
south of the country. It operates from approximately 30 locations, from
Leicester and Norwich down to Winchester and Crawley. In 1991 the
company operated nine landfill sites around the region. Hales have
around 500 vehicles for shifting waste to these and other sites. The company disposes of domestic waste for local authorities and industrial waste
under contract to a variety of companies. Urban waste is transported to
transfer stations and taken from there to the landfill sites. These are

usually close to the transfer depots in order to minimize transport costs. However, the difficulty of finding suitable sites means that transport costs are increasing. According to a landfill manager at Hales, commenting on the Bierton proposal: "At the moment we are using worked out quarries, both our own and other companies. But we are looking to a land raising exercise. Basically this is a greenfield site, low-lying farmland or land that will lend itself to land raising . . . [There is] a shortage of quarries, a shortage of holes in the ground. Everybody makes waste, it's got to go somewhere". The shortage of worked out quarries is exacerbated by the opposition to landfill. As the same respondent put it: "[We] are very conscious of doing things the right way to say the least. But at every step we take we get knocked no matter what we do. Really we can only liken ourselves to the old early-nineteenth-century scavengers. Its a pretty basic job we do and everybody looks in the other direction". Consequently: "you are always seeking land for landfill".

While the industry is short of sites, the problems afflicting agriculture are driving many landowners to landfill as a viable option for their land. The negotiations between Hales and the landowner are complex and subject to a whole variety of conditions. The landfill manager explained it thus:

> We'll have a look at the land, make an assessment of what we think we can do with it and very often we just come up with an option deal initially. We broadly talk about money, whether they want it up front or on a revenue or royalty basis. So much depends on their tax situation and we work It accordingly. . . . If it's a site with good road access and there's a good chance of us getting planning permission, a long-term site, everything is fine and we take into account our approximate site preparation, total volume we'd like to put in there and the standard the farmer would like the land brought back to and our ongoing obligations which would be 50 years after the site's finished. All these things have to be taken into account. If it is a site which is in a remote country area, it's going to be long term and the type of waste we put there is largely builders' waste. We'd be probably offering, I don't know, 50p per cubic metre of air space in the ground . . . into that we can put two [cubic metres] of waste.

In the case outlined here, the farmer of the land, being a tenant, had little say over the proposed development as Hales negotiated directly with the landowner (an insurance company).

The proposal

Hales operated a site nearby, but this had little spare capacity so the company was looking for another site in mid-Buckinghamshire. This led them to Bierton and the proposal to raise a parcel of agricultural land lying just outside the village boundaries. Essentially the objective was to re-contour the area of pasture using waste materials, thereby raising the level of the land by a maximum of 5 m. This would involve the importation of 950,000 cubic metres of waste material, half of which would be industrial and commercial waste, with a small proportion of household refuse. It was proposed to tip the waste material within five phased areas which would be subdivided into cells. Each cell would be approximately 50 m wide and separated from adjacent cells by 5 m bunds following the lines of existing hedgerows, which would be retained. There would be a total of 16 cells and each would have the capacity to import waste for six to nine months. The cells would be constructed by first removing the topsoil, which would be stored. When each was filled with waste it would be capped to a depth of 1 m using clay extracted from the next cell being formed. Sub-soils and topsoils would then be replaced to a total depth of 1 m. The restored land would be grass seeded and hedges, where removed, would be replaced.

According to a landfill manager the object of this operation was to minimize the area tipped on at any one time: "People think if you've got a 300 acre site you are going to tip all over it. You are only tipping small areas; none of this 'start work here and work your way across then restore it'. In a modern landfill site you bund it, you line it, you do a section at a time and you restore very close behind. You look at any one of our sites and you can see the preparation and restoration. All there as we're going along because we don't want to keep the land open". On this site it was proposed that the operations would last for nine years. There would be, on average, 120 daily vehicle movements, with a maximum of 200 per day.

The development process: the actors and their representations
The publication of this proposal brought an almost immediate response from the nearby villages, particularly Bierton. On 8 March 1990, the *Bucks Herald* carried a major article on the proposal. According to the report:

> Villagers in Bierton are preparing to fight plans for a giant waste tip which they say will be an ecological disaster for the entire north side of Aylesbury Vale. . . . Villagers have been meeting over the past week to form an action group to oppose the plans, which they say will cause traffic congestion and knock the bottom out of the area's property market. After one meeting the group issued a statement saying the prospect of 120 lorry movements every day, seven days a week for nine years, bringing rubbish to Bierton and Hulcott's doorstep was totally unacceptable." (*Bucks Herald*, 8 March 1990)

According to an early participant in this action group he first became alerted to the proposed tip when a neighbour knocked on his door in early March. They then spoke to the farmer of the land and the Leader of the Parish Council. Having established that the proposal was due to be considered by the Parish Council at the end of the month, they then studied the development plans to get some idea of the implications. By the time the proposal came before the Parish Council almost the whole village was aware of what was happening and the meeting was full. At this meeting the Parish Council sanctioned the formation of an action group. According to one participant: "There was already an A418 action group in existence [concerned with a proposed bypass] but we wanted a separate group, one which was concerned only with this proposal and would be disbanded when this was over. I had been in an action group previously when we lived [elsewhere] . . . that had got political. I was determined that this one should not get political. So we restricted ourselves to this one issue". The local vicar joined the group and allowed the church hall to be used for campaign meetings. In early April a meeting was attended by more than 600 people. They came from all the villages surrounding the proposed site. Two county councillors were present (including the Chair of the County Council) and a representative of the local MP (George Walden) who said: "Mr Walden is very concerned about the environmental impact of the proposal . . . He is sure it will be a very successful fight – it has to be" (*Bucks Herald,* 12 April 1990).

At this meeting an action group committee was formally proposed and accepted. According to the *Bucks Herald:* "the action group who were formally elected on Monday have been working together in private since news of the plans reached the village" (12 April 1990). This was borne out by one of those involved, who said:

> At that meeting a committee was formed. Now this was not done democratically, a group of us had met beforehand and formed the committee so at the meeting we proposed these people, this committee, and it was accepted. It included representatives from all over the village. We wanted the whole village included, not just people down this end near the proposed site. Mostly professional people with some expertise in the position they've taken: we have a banker as treasurer, our press officer is being advised by the Bucks County Council Press Officer (who also lives in the village). Lord Alexander, Chairman of NatWest Bank, is our President. His house unfortunately looks right across where the proposed tip will go.

The rôle of Lord Alexander attracted national press coverage. *The Times* quoted his Lordship as saying: "I share the deep concern of villagers at this proposal to despoil this precious area of rural England. It is only necessary to drive from the Midlands through Milton Keynes towards Aylesbury to realize just how fragile is the area of countryside which divides Greater London from the Midlands." (23 April 1990).

The action group committee began by meeting fortnightly, either in the local school or at each other's houses. The first action was a letter to the local councils which stated:

> We represent the people of Bierton, Hulcott, Rowsham, Wingrave, Aston Abbots and Weedon [the villages surrounding the proposal site] in expressing our grave concern at the despoliation of this beautiful part of the Aylesbury Vale . . . . We object vehemently to the proposal, believing it to be grossly ill-judged and an affront to all who value and seek to conserve this area of unspoilt countryside which still fulfils its centuries-old rôle as valuable grazing and water meadows at the heart of a productive farm".

The bulk of the letter dealt with the proposal in terms of its contravention of Structure Plan policies. First, permitting "waste disposal operations on this site would blight the delightful rural character of this part of the Vale and form an ugly blot on the tranquil and unspoilt rural scene . . . . The site would clearly be a clearly visible intrusion to the circle of neighbouring villages which surround it". Secondly, the application "would clearly fragment the farm and put its very future at risk". Thirdly, "a waste disposal operation of this kind would not 'restore the land to contours resembling the original topography or to contours which would enhance the scenic value of the area'". Fourthly, "it is highly doubtful if the applicants, however well intentioned, could meet the requirements of paragraph 51 of the County Structure Plan 'to protect watercourses and aquifers, residential or recreational areas from pollution . . . sites are to be refilled with waste materials'". Fifthly, "the environmental effect [of the tip's lorries] . . . would be traumatic. It would also fly in the face of Government claims that their policies are designed to relieve small communities of the noise, pollution and danger caused by heavy lorries passing close to their homes in totally unsuitable narrow streets". Finally, the application would not meet the requirements of Structure Plan paragraph 55 which states that: "Where there is no economic alternative to tipping . . . it is the policy of the County Council to ensure that tipping takes place where the usefulness of the land for agriculture, recreation and forestry or other appropriate use, can thereby be improved within the shortest practicable time". In this case, the group argued:

> There are economic alternatives to tipping at this site. There are disused brick-pits in Buckinghamshire . . . which are already accepting vast quantities of waste and which have a working life of many years. . . . In any event, there must be many other areas in Bucking-

hamshire and neighbouring Hertfordshire where waste tipping would be more environmentally acceptable than on an unspoilt, agriculturally productive site with watermeadows in one of the most attractive stretches of countryside in the Vale of Aylesbury.

This document is of interest because it shows that the action group attempted, from the very beginning, to structure their arguments around development plan policy in a very professional fashion. There were two main dimensions to this: first, the effect on the local environment and secondly, the question of "need". Effectively, the action group argued that the environment of the proposed site was totally unsuitable for tipping and there were in any case sites elsewhere much better suited to landfill.

As well as this "policy" response the action group mobilized the villagers in demonstrations of their opposition. The village was plastered with "No Tip Here" posters. When the District Council planning committee considered the application in late May the meeting was picketed by "more than fifty Bierton villagers . . . chanting 'No Tip Here', 'Save the Vale' . . ." (*Bucks Herald*, May 24 1990). The District Council had, by the time of the meeting, received 375 letters of opposition, as well as official protests from parish councillors at Bierton, Wing, Hulcott, Wingrave, Aston Abbots and Weedon.

At this meeting the committee "unanimously condemned the controversial proposal" with one Councillor calling it "the most terrible in terms of damage to the environment that had ever come before the committee" (*Bucks Herald*, 24 May 1990). The District agreed to inform the County Council of their total opposition to the plan. This opposition was on the following grounds: (a) the proposal would not result in the improvement of the land for agriculture or conserve and enhance the scenic beauty and wildlife interest of the adjoining Area of Attractive Landscape; (b) it would seriously prejudice the high level of visual amenity which exists in this rural location; (c) the roads leading to the site are inadequate; (d) the applicant has not satisfactorily demonstrated that leachate would not migrate into the water table seriously polluting the surrounding area in the longer term; (e) no environmental impact assessment has been carried out as required under the 1988 regulations; (f) an accurate ecological survey of the area should be undertaken; and (g) the existing landfill facilities in Aylesbury Vale District provide adequate capacity for existing and future landfill requirements (AVDC Recommendation to the County Council). These objections were very similar to those raised by the action group in its letter to the District Council.

Having learned that the proposal fell within Schedule Two of the Town and Country Planning (Assessment of Environmental Effects) Regulations 1988, which stipulated that an environmental assessment of the proposed site should be made, the action group decided to formulate their submission to the County Council in that form. According to an action group member: "We approached the whole thing on the basis of an environmental assessment. This took us into the area of technical detail. So we needed to see who we had available locally to deal with these technical issues". The group drew together local "experts" ("the quiet wealth of Buckinghamshire" as one respondent put it) including a consultant geologist, a retired public health officer, two local ornithologists, a member of Shell's agricultural division and an ex-member of the design team for the M25 motorway. There was a division of labour in the group, whereby technical teams were established to deal with each section of the report. Each compiled its own section and all were finally collated. The group believed that the application would be heard on 1 October, so they worked hard through July and August to get the report completed and circulated to members of the planning committee.

The report itself was prefaced by a brief biography of each contributor and had seven sections. These were: (a) a letter from the Chairman outlining the group's objections; (b) an examination of the proposal from the perspective of the Structure Plan (akin to the letter referred to above); (c) a consideration of the implications for water pollution, land drainage and soil; (d) a review of the implications for environmental health; (e) an examination of the effects on flora and fauna; (f) a consideration of the likely impact on agriculture and air pollution; and (g) an assessment of the effects of traffic and the implications for the road system. Also included was the assessment of the site made by the County Ecologist and a local representative of BBONT, who had found 44 flowering plants, mallard ducks and "clear and rich" brook water on the upper boundary of the site. The manner in which this report was produced shows the level of human resources available to the action group in representing their case. The report was word-processed by a former head teacher of a business studies centre and a professional photographer was recruited to provide illustrative material. Seventy copies were produced through a "local businessman" in the group. They took the decision to produce *their own* environmental assessment and in so doing were able to draw upon a significant level of technical expertise. They were able to produce an assessment which, if produced by consultants, would be extremely expensive.

Although the action group had prepared this report for the County Council Planning Committee meeting in October it soon became clear that the application would not be heard at that time. The decision date was pushed back to later in the year. Nevertheless the action group tried to mobilize the villagers for the meeting whenever it might come. A newsletter in September argued that "we must show the County Council – and Hales Waste – the great strength of our opposition by turning up in force at County Hall by at least 10am on the decision day. Your Action

Group will, of course, let you know in good time whether it will be November 12th or December 17th". The reason for the delay was Hales's lack of response to the planning department's request for an environmental assessment. According to a planning officer the County were keen on acquiring an assessment because they wanted full information on the site itself, and in any case such an assessment was a statutory requirement under Section 2 of the 1988 Town and Country Planning (Assessment of Environmental Effects) Act .

On the other crucial issue, that of the "need" for the site, the company approached the County Engineer for advice. The County Engineer submitted his "advice" to the planning department in early November and showed that the question of "need" was far from clear, because the relationship between demand and supply is not, as in the case of minerals, structured within the formal planning process. For instance, the County Engineer concluded a letter to the County Council by saying: "I am sorry I cannot supply you with 'hard data' and facts but I believe the 'needs' question is complex and is not solely the strategic one of balancing total waste to available void, but the more local one of transportation and other factors of availability, competition etc." (Buckinghamshire County Council Planning File). Given the complexity and difficulties surrounding the question of a "need" for the Bierton site, the decision-making criteria became fully focused on the site itself.

The problem was that the environmental assessment by Hales was not forthcoming. The company wrote to the Planning Department in November claiming that their consultants had undertaken studies of the site which would lead to only minor modifications in the proposal (mainly to take account of the County Ecologist's report). The delay in providing the assessment was blamed upon "the exceptional weather conditions during the summer and early autumn of this year, the resulting hard ground conditions have made the implementation of important investigations impossible until recently". Another letter from the company, a week later, reiterated the point that the Committee should wait for the statement: "It is of some concern to me that there appears to be a lack of understanding of the depth and scope of work involved in the preparation of an Environmental Statement".

As the year ended no Statement had appeared and the decision was made to hear the application in February 1991, when the proposal would be one year old. For this meeting the action group mustered around 150 demonstrators (*Bucks Herald,* 7 February 1991), and the Planning Committee heard that 413 letters of objection had been received. The application was refused on the following grounds:

- it would constitute unwarranted development in the open countryside
- the proposed tip would not lead to any improvement in the

usefulness of the site for agriculture, forestry or recreation
- the development would result in the loss of an area of land of ecological value
- the tip would be potentially damaging to wildlife
- the development, if permitted, would give rise to an increase in the number of vehicles using local roads.

The Planning Committee was unanimous in its opposition to the scheme.

The scale of opposition to this proposal was extensive. The local action group ran a professional, well organized campaign. Not only were they headed by a figure of national stature but they also commanded significant cultural and economic resources in making their representations to the planning authorities. They presented their case in terms which attempted to identify the key structure-planning implications of the proposed development. However, they were able to go further than this; in undertaking an environmental assessment they were attempting to match a possibly favourable assessment from the developers.

This case indicates that the requirement for environmental statements or assessments in cases such as this will open up another terrain of contestation. There seems to be scope for this within the regulations; the DOE Circular offering advice on implementation of the 1988 Act states ". . . the process of assessment will involve consideration of environmental information from a number of sources – from the developer, from statutory consultees and from other third parties as well" (DOE Circular 15/88; see also Mertz 1992: 497). The result of this process may be to add to the inequality between places and the ability of communities to resist the power of developers.

In this case the developer seemed only to make a half-hearted attempt to push the development through. Hales seemed to have made little effort, in comparison with the action group, to persuade either the Planning Authority or the local community that the development was either necessary or beneficial. Hales's attitude to the local community is indicated in this comment from a landfill manager: "Whenever we put an application in, as soon as that notice goes up on a gate, we get objections. Some are really well organized but I can assure you we will have objections at every stage. We try and talk to people and sometimes they won't listen, other times they will. And I've been to meetings and I've stood up and I've been attacked for two hours solid because the chairman wasn't a neutral chairman". The company clearly expects a hostile response to landfill applications and may well have concluded that there was little point trying to win over the local community. However, they may also have felt undermined by the environmental assessment drawn up by the action group, and this may account for their own inability to get an assessment done. This points to the potential power of action groups in places such as Bierton, with its abundant stock of cultural and economic "capital", for they may continue to fend off development while other places are less able to do so and therefore find themselves on the receiving end of unwelcome development. There was a glimpse of this in the above case as the Bierton action group argued that other sites in the county, where there was spare capacity in existing landfill operations, were much better suited to waste disposal. Clearly, places with existing landfill operations stand a good chance of having these extended. One such site was close to the village of Newton Longville where landfill had been going on for many years. This village was included in our social survey which showed that 72.8% of the village respondents wanted less waste dumping in the countryside, with 43.5% of these wanting none at all. However, despite this level of antagonism to the industry these villagers will have to live with landfill for some time to come because of the difficulty in finding new sites.

Back to the context

The shortage of available waste disposal sites drove Hales to make what seems, in retrospect, an inappropriate proposal. The way this matter was dealt with gives some indication of how this shortage has come about. There is clearly local animosity to landfill operations and the planning system contains plenty of scope for this to be mobilized, particularly where local knowledge of the issues can be effectively represented within the planning process. The wider issue of waste disposal in the regional context was not adequately raised in this case. The "needs" argument, half-heartedly put forward by the County Engineer, contained very little in the way of strategic (non-local) considerations. The absence of a Waste Disposal Strategy at the county level was partially responsible for this. The government has now issued draft proposals for the preparation of waste disposal development plans and the County Council is currently in the process of drawing up a waste disposal management plan with the broad aim of identifying the amount of waste to be deposited in the county and the sorts of areas in which this might be accommodated. These would be similar to local minerals plans and would provide "the framework within which individual planning applications should be considered. Without a requirement for waste disposal development plans, there is a possibility that waste disposal could be the only land-use topic for which there is no development plan prepared by the relevant development control authority" (*Planning* 893, 2 November 1990: 23).

In response to these proposals the National Association of Waste Disposal Contractors (NAWDC) expressed concern that the preparation of local plans goes no way towards identifying regional needs: "There is increasing evidence that county planning authorities are seeking in their structure plans to adopt policies relating only to wastes generated locally. The Association believes that waste management development plans will eventually have to be tackled on a regional basis" (*Planning* 900, 11 January 1991: 25). This response is attributable to the economic and technological changes taking place within the industry as well as the tighter regulations enshrined in the Environmental Protection Act. The tendency is for landfill and other disposal operations to become "fewer and larger, thus needing a wide catchment area to pay their way" (*Planning* 900, 11 January 1991: 25). It is estimated that profits on landfill sites are on average 50% while on cheaper sites the margins may be up to 70%; so while the industry is becoming more tightly regulated it is also becoming more profitable (*Financial Times*, 26 November 1991).

However, the resistance to landfill at the local level is making the industry look to regional planning as a means of strengthening its hand in individual planning cases. This can be likened to the minerals case study examined earlier, where the regional apportionment of a quota to the County Council almost ensured that permission for development was granted in a case which was equally locally unpopular as the Bierton case. The waste disposal industry's support for regional planning can be seen in this context. According to the technical director at Shanks & MacEwan, one of the largest waste disposal companies: "The rules are okay. It's the enforcement that's the problem. It's so often being done at the wrong level. We need common standards. It would be very useful to see regional bodies, like the National Rivers Authority introduced for waste disposal" (*Financial Times*, 26 November 1991). This view seems to have found favour with the government who have suggested the inclusion of waste regulation in the proposed National Environmental Protection Agency which would thus deprive local authorities of their responsibilities in this field. Not surprisingly the industry supports the proposal. The policy executive at the NAWDC said: "We want local authorities taken out of the picture. We want national standards and we want them strictly enforced". The Association of County Councils has opposed this move and in its representations to government has stressed that the concerns of local people cannot be met by "a remote national agency" (*Financial Times*, 26 November 1991).

Conclusion

The case studies on minerals and waste disposal presented in the previous two chapters illustrate large-scale and site-specific development proposals which failed. In one sense they failed because they are the types of activities that most influential rural residents in Buckinghamshire do not want to see in their neighbourhoods. However, before we present a potentially misleading picture of development patterns in the county, we should say that this does not mean that *all* such proposals will fail. Clearly, minerals will continue to be extracted and waste will be dumped. In this sense our case studies are not representative of *all* such proposals in the county. They show special characteristics. As we have emphasized, the villages affected by these proposals clearly demonstrated that they have great powers of resistance based upon the levels of "professional" resources which reside there. In this sense we believe that these cases are representative of a growing trend. Minerals extraction and waste dumping *are* becoming more difficult in areas such as Buckinghamshire, and this is evident in the exasperation of the two development companies involved in these proposals.

What is also crucial here is the rôle of the planning system. In this aspect the two cases are distinct. In the first case a strong sectoral planning framework was in place which attempted to ensure that some minerals, on a scale roughly in line with the national and regional targets, are extracted from particular areas. Thus, the opponents of specific proposals find themselves faced with an elaborate machinery whose purpose is to assess the "need" for aggregates. In waste disposal, however, and despite the efforts of the industry to acquire such a framework, the assessment of "needs" is open to more local definitions. As the case study in this chapter shows, the needs issue sometimes may be extremely unclear, puting the site itself at the centre of the contest. Waste disposal companies have therefore to come up with good sites, a task which is clearly becoming increasingly difficult. Hence the mobilization of local residents against a waste proposal can be extremely effective, especially where there are considerable resources available to the local community. In the case study presented here there was a lot of professional expertise available locally, including prior involvement in other action groups and access to technical resources, exemplified in the presentation of a "home-made" environmental assessment. This "professional" approach (a hallmark of "professional people") allowed easy comprehension of the complexities of the planning system, turning this form of representation into a resource in the local campaign against the development (compare this to the developer who saw the planning system as simply an obstacle). The other resource which was clearly drawn upon was the "unspoilt" character of the local environment. There is a general point to be made here: the more unspoilt the environment the more weight it will carry in the negotiations over planning permission. It can much more readily be constructed within the development discourse by those seeking to prevent development than by those who wish to alter it in some way. In the above case it was very difficult for the developer to show that the landscape would be any better after the development than it was before (indeed ADAS foresaw only a marginal improvement in the agricultural potential of the site). Why then endure nine years of landfill, especially when the need for the site is far from proven? Places which constitute "nice" environments are likely to attract those with sufficient resources to live within such expensive and exclusive spaces. They are precisely the people with the resources capable of defending those environments. But if development has to take place somewhere, then where? Who gets the minerals workings and landfill sites of the future? The short answer would seem to be, those with the powers of least resistance.

CHAPTER 9

Industrial development and the limits to growth

Introduction

In Chapters 7 and 8 we looked at the rôle of middle-class action groups in preventing unwanted development in their immediate localities. What was common to these two cases was that both developments impinged overtly on settlements and thus mobilized opposition from large numbers of residents. However, much of the diversification of the rural economy takes place on farms in more low-key developments than those we have so far examined and clearly not all development proposals lead to the formation of action groups. This is not to say, however, that the influence of middle-class residents is exercised merely through direct action. As we saw in Chapter 1, middle-class groups have generally been regarded as very adept at influencing the forward plan-making process. We do not intend to present here an analysis of middle-class involvement in development plan making; instead we wish to focus on the ways in which putting the plans into operation can lead to a curtailment of certain (industrial) activities which are not deemed fit for the new rural spaces. We will show how certain actors in rural areas find their attempts to diversify the economy and their room for manoeuvre, limited by the implementation of development plans.

As we outlined in Chapter 5, during the 1980s attempts were made to curtail the scope of rural planning in an attempt to promote rural enterprise. By the end of the decade this had actually resulted in a strengthening of the planning system. We here chart this shift by focusing upon the development of small-scale industrial units on farms in Aylesbury Vale. In doing so we attempt to bring together three contexts: (a) the changing nature of the labour market; (b) the decline in agricultural productivism; and (c) the methods employed by the land-use planning system to regulate development. The strategies of actors in the land-development process (i.e. in this case the attempts to develop small-scale industrial units on farms) are assessed in relation to these changing contexts. The analysis thus represents an attempt to focus upon the transition of economic activities (from agricultural to industrial) within the study area.

The encouragement of exemptions to the *general presumption against development* offered to rural planning authorities during the 1980s by the DOE and MAFF through a series of circulars, planning guidance notes, grants, etc., was an attempt to assist private-sector developers (e.g. farmers) in their negotiations with planning authorities over individual applications for development. Moreover, changes in the Use Classes Order (1987) and General Development Order (1988) also significantly altered the *transfer rules* concerning the use of industrial and commercial space. In particular, the broadening of the B1 Business class to incorporate both light industrial and commercial activities was significant for a number of reasons. It embraced a variety of uses (giving more flexibility for developers to modify use) which formerly constituted separate planning uses requiring specific planning permission. Also, premises in the B2 General Industry use class could be changed to B1 use without restriction. As a result it was widely believed that more industrial space would come into the light industrial and office sector market.

In the SE region as a whole the demand for industrial and commercial space showed a rapid increase during the 1980s. This demand was less buoyant in the provision of traditional industrial space but was most marked in the B1 use category (i.e. office and light industrial developments), leading to rents increasing by approximately 35% in the latter part of the decade (before the onset of recession in 1990). Ninety-five per cent of the UK's high tech floorspace was located in the SE region, with 80% of this in the ROSE area. Given the overall regional constraints on the supply of industrial floorspace (particularly in the B1 category), the opportunities opened up by the planned expansion of supply in Milton Keynes contributed to the area of north Buckinghamshire being recognized as the fastest growing area of development in the UK (see Champion & Green 1988). Milton Keynes experienced strong growth in service employment and relatively low levels of unemployment. During the late 1980s in particular, the city was in receipt of high levels of capital investment from the private sector. For instance, in 1988–89 a further 117,970sq.m of floorspace (105,334sq.m industrial, 6,395 for offices and 5,827 for retailing) were added to the existing stock within the designated area. Nevertheless, rents for industrial and commercial floorspace increased by 25% in 1987–88 alone. The pressure for new floorspace in the SE during this period is also evident in the amount of agricultural diversification in this sector. Empirical evidence suggests that the growth of industrial units on farms between 1985 and 1989 was greatest in this region (see Kneale et al. 1991).

It is in this context of decentralized and organic economic growth, together with variable levels of planning restraint in rural parts of the county during this period, that we have to view the development of industrial premises on farms in Aylesbury Vale. Given the growth in demand for small types of industrial space units and the high levels of rent for space in established centres, it seemed natural that farmers with redundant agricultural buildings should begin to think about converting them to other uses, particularly as DOE and MAFF seemed to think high tech and certain types of service industries could be appropriately sited in rural areas. Thus, in Aylesbury Vale, farmers attempted to meet the new demands for industrial and commercial floorspace, particularly in the B1 category, i.e. light industrial/commercial space. The relatively high rent levels in Milton Keynes and Aylesbury encouraged demands for cheaper more flexible space, especially for small businesses, in other parts of the locality.

In order to show how farmers diversified their farm building uses and how this brought them into contact with a planning system that was in a state of flux we present here a series of case studies of industrial development on farms. The case studies follow a particular time sequence. They start during 1987 and conclude with a case in September 1990. The time period of development and negotiation was crucial for the extent to which the land-development process was successfully completed. As we have made clear, the period 1987–90 represented a time of adjustment in both local planning and national policy in relation to rural land development. The case studies illustrate this process of change in the regulatory procedures and explore the changing rôles of the key actors and the contexts in which they operated.

Rural industrial development case studies

Case study 1: Mr H., 1987–90

Mr H. succeeded to an 88 ha farm on his father's death in 1978. Death duties meant selling all the land except for 16 ha. The principal enterprise was pigs. Owing to poor profitability these were sold off. Mr H. then developed a full-time contracting business, putting his 16 ha down to arable crops (wheat and oil-seed rape). The general instabilities in agricultural markets, together with the continuing obsolescence of his contracting machinery, led to the sale of the contracting business in 1986. He started looking around for another activity and considered the traditional diversifications – crafts, holiday lets etc. – but they seemed unable to provide a stable guaranteed all year round income. He contacted the Rural Development Commission (RDC) and discussed his problems with the local area officer. The officer suggested that he advertise small industrial units in the local press to test the level of demand. He received 10 or 12 replies. In this way he discovered that there was a need for industrial units in the countryside.

Mr H. then looked into the kinds of units provided by the councils and by Milton Keynes Development Corporation that were available in the towns. They tended to have quite large units but nothing in the 400–700 sq. ft range. With the help of the RDC Mr H. applied for planning permission. He quite assiduously gathered the support of key people such as local NFU representatives, the local County Councillor and the Parish Council. According to Mr H. the planners were quite suspicious, seeing this as an attempt to convert redundant farm buildings into a rural industrial estate. They saw this as particularly problematic as they were providing industrial floorspace in the towns. Nevertheless, Mr H. lobbied hard and assembled a well thought out package. He was also the first farmer in the district to come forward with a proposal of this kind. He was granted planning permission without any real problems.

According to Mr H. the planners were quite keen that he had traditional craft- and agriculturally-related industries on the estate. But he realized that they would not be able to afford the kinds of rents he would need to charge to make the enterprise pay. He also did not want any heavy industry, mainly because of the ensuing traffic problems. He widened an access road to cope with the increase in light vehicles and was reluctant to create further problems on the small country roads which lead to his farm. As a result of these considerations, and on the advice he received, he decided to accommodate only Class B1 light industry, and designed the units for this purpose.

The problems surrounding the provision of the first set of units were the most serious. Capital was needed to provide the facilities to service the first units – these included sewerage, electricity and water. Having made that initial outlay he then had the facilities to service a further ten units. The initial units meant a conversion cost of £20–40 per sq. ft but that was reduced to an average of £15–20 for the rest. With the rent set at £5 per sq. ft there was a four to five year pay-back period.

Mr H. provided ten units ranging from 400–2000 sq. ft. He then rented these out to most of the people who had answered his advert. These included: a kitchen furniture maker; an import/export business; a computer software firm; kits for racing cars; a prefabrication engineer; an electronic gate manufacturer; industrial cladding; a music publisher; and electronics manufacture. Around 70 people were employed on the estate.

There seemed to be a variety of reasons why these small businesses wished to move out into the countryside. Mr H. believed that one of the most important was the length of the leases on offer. The local authorities prefer long leases – 21 or 25 years. But, as Mr H. put it "half these young fellas don't know what they are doing for the next 25 months never mind the next 25 years". Mr H. offered leases of 3 and 5 years and tried to be flexible if someone wanted to leave – they had to give him a year's notice. Balanced against this he reserved the right not to renew the lease at the end of the rental period if he so wished. He claimed that this had not been necessary, although he had not been in the business long enough to know how necessary it might become, particularly given the onset of recession when many small firms might be at risk.

One of the other main reasons he cited for the popularity of the units was rent levels which were set marginally below local authority rates. Also important was the preference of many small firms to move out of the towns to the "peace and tranquillity" of the countryside. They had a good working environment and none of the traffic problems associated with travelling into congested urban centres.

Mr H. laid great emphasis on the point that he was bringing employment back to the countryside. He argued that the mechanization of agriculture, the rationalization of service provision and the decline of non-agricultural employment in the countryside all contributed to a diminution of the rural workforce. The influx of many middle-class people who worked in the towns to live in the countryside meant that many of the villages in the area were virtually deserted during the day. Thus, 70 people employed on one farm made a significant difference to the custom of local shops, petrol stations and pubs. There were therefore spin-off benefits for other local businesses in the area.

Mr H. was one of the first to initiate this activity in the locality and he capitalized on it by opening up a consultancy, advising other farmers on how to establish industrial units. About 10% of his income came from this source. Other farmers seeing his success attempted to follow suit. One such was a friend of Mr H. (Mr T. – case study 2) who ran a nearby pig farm.

Case study 2: Mr T., 1987–90

This farmer attempted to set up industrial units in 1987. He'd previously been a pig farmer with 120 breeding sows. At that time pigs had become extremely unprofitable and he decided to cut his losses and leave the industry. He did not decide immediately on industrial units, looking first at a variety of fairly obscure enterprises, all, incidentally, associated with animal production of one sort or another (they included worm farming, snail production and trout, sheep and calf farming). None of these seemed to provide the sort of income that he was looking for. He consulted a local agent to help him to decide what to do and they eventually decided upon industrial units. Although the agents had no previous experience of these the RDC officer gave the same advice as to Mr H. The agent and the RDC officer helped Mr T. draw up his plans. Mr T. said that at that time the planners were very sympathetic. The only real problem lay with the Highways Department over entrances to and exits from the site. Once again the RDC officer played an intermediate rôle, successfully negotiating the planning agreements.

It took six months to obtain planning permission after which Mr T. began to convert the buildings, employing electricians, plumbers and plasterers as needed and it took four months to rip out the interiors. He initially built one unit as an exhibition unit but, as in the previous case, he then had to provide the facilities for all ten. This involved constructing sewage treatment tanks, a toilet block and a car park. Mr T. did not even bother to advertise. Before he had completed the first unit a high tech computer company leased it and this pattern was repeated. As the units were finished they would be occupied immediately. The occupation of the units occurred by word of mouth.

Mr T. wished to maintain the tranquillity of the farm. He did not want industrial activities in the units as they would produce noise, mess or toxic substances. Although he had enquiries from car mechanics, carpenters and the like, he refused them, preferring "clean", high tech industries. He finished up with two computer companies, an office studio, a kitchen and bathroom manufacturer, a picture-framing company, a graphic designer, a cabinet maker, a curtain (soft furnishing) manufacturer, a printer and a pine furniture company.

Mr T. gave almost exactly the same reasons as Mr H. for these firms wanting to be in the countryside and listed exactly the same benefits. He did, however, seem to be having more problems adapting to his new rôle. He had been a pig farmer for 15 years and admitted that he was having problems making the shift from "managing pigs to managing people". It had taken him a long time to get accustomed to having people around on the farm as it had previously been his domain. He was undertaking all the building work himself and this kept him occupied over the first two years. He had converted 11 units and had 6 still to develop. He expressed some concern about what he was going to do when the building work was completed. He would simply be an "estate manager" and couldn't see exactly what there would be for him to do. Having been in essence a manual worker all his life, this was worrying for him.

Both of these cases outline the advantages of having industrial units on farms and both introduced them before they became popular in the locality. This involved a significant amount of financial risk, particularly in the early stages of the development process. They were supported and encouraged by external advisers, and the local planning system saw no reason to reject the applications. However, the planners began to tighten up planning procedures in such cases. Mr T. reckoned he would have had difficulty in getting his proposals through had he applied two years later. We will now turn briefly to a proposal which encountered some difficulty obtaining permission at a later date.

Case study 3: Mr M., 1988–90

This farmer submitted an application in 1988 for two barns to be converted into class B1 light industrial premises. By this time the planners had begun to reconsider their approach to this type of proposal. Mr M.

thus encountered problems due to the non-traditional nature of his farm buildings. However, for various reasons this proposal took a long time in coming before the planning committee, only reaching that stage in early 1990. The proposal was initially passed on the grounds that the applicant had waited an inordinately long time for the outcome, although there was speculation from one of our respondents that certain members of the committee had been "spoken to behind the scenes". However, the planning department opposed the application on the grounds that these buildings were not suitable for this particular change of use. The only real reason for reaching this conclusion was that they were new buildings. Having now seen a number of these proposals being passed the planning department was increasingly looking for a way of restricting the growth of industrial units in the countryside. One way of doing this was to insist that only proposals in "traditional" buildings should be considered. They believed that the industrialization of agriculture resulted in buildings being constructed which, had they been subject to planning control, would never have been permitted. They now saw the shift of farmers into new enterprises and their subsequent need for planning permission for many of these, as an opportunity to control this process "in the local interest". At the very least they did not wish to see industrial buildings in the countryside full of "urban" industries.

Case study 4: Mr A., 1990

Mr A. ran a 204 ha arable holding with his brother. After succeeding to the farm (in 1959) the partners had shifted enterprises, moving from dairying to beef and arable farming. In 1976 they established one of the first "Pick Your Own" enterprises on a 8 ha site near the farm, growing and selling 17 different kinds of fruit and vegetables. A lake was also excavated to provide irrigation. This was filled with trout and advertised as a trout fishing site. The beef enterprise was withdrawn in 1983. Mr A. and his wife started a bed and breakfast business. By late 1989 the diversification strategy was unable to subsidize the losses accruing to the arable enterprise. The established "Pick Your Own" enterprise had suffered over the previous five years owing, he argued, to the more easy availability of fruit and vegetables in the retail chains. Financial losses were in the order of £20,000 per year and it was now believed that the farm could no longer support two households. The strategy was to disengage from farm production and to realize assets in a number of ways, while at the same time maintaining family occupancy of the farm. At the time of interview a series of different *asset realization* strategies were unfolding. These concerned: (a) selling off 8 ha of land for paddocks; (b) attempting to sell most of the rest of the land as agricultural land, excluding 4 ha surrounding the farmstead; (c) converting many purpose-built farm buildings to industrial use and letting these onto the market; and (d) maintaining residence in the farm house and some of the adjoining buildings.

Such varied asset realization strategies required participation in a variety of advice networks. These included: (a) *land agents*, who were important in advising on the sale of his agricultural land. They were advising him to set aside most of it in the interim period because of the slack market; (b) *estate agents* (a local long-established firm) who were handling the sale of the paddock land that had been on the market for at least a year; (c) *industrial unit consultants* (Mr H. above and the RDC) who were advising on the potential for converting farm buildings. These consultations had not been a success. Mr A. thought that it was better if he adopted a "go it alone" strategy. He advertised in the local paper for prospective tenants, immediately receiving 15 replies; (d) *planning authorities*, seen as very much the "dead hand" and as inconsistent in their decision-making. His past attempts to obtain planning permission had failed. As a result he thought it best to develop "industrial units first and face the consequences afterwards".

The conversion of farm buildings to industrial units was seen as economically viable but there was considerable reluctance to provide facilities, undertake structural changes to the buildings, and generally pursue a more formal process of development. Mr A. regarded the knowledge gained from his social networks (concerning the higher level of rents for industrial premises in urban areas and the potential gains in revenue from setting up units) as sufficient.

Mr A.'s informal and somewhat multi-dimensional approach to asset realization became caught up with the downturn in the economy, high interest rates and, as we shall see from the final case study, a local rural planning system which was to develop a more coherent strategy vis-a-vis building conversions. As a result the viability of his strategy was brought into question. He was, however, forced to adopt it. Consultations with bank managers and accountants were driving this course of action. With £70,000 in outgoings and only £50,000 income, overdraft facilities were being extended while land was used as collateral.With the potential of converting 20,000 sq. ft. of redundant agricultural floor space to industrial uses, Mr A. had little option over the next two years but to pursue the conversion process.

These case studies illustrate what is entailed as farmers disengage from agriculture and seek to establish new enterprises on their farms. They must first identify a way of reutilizing their assets, both buildings and labour. Often this is done by using advisers and agents, principally the Rural Development Commission. Having chosen a strategy the planning system must be negotiated, again using external advisers. To be successful, as the first case study shows, it is helpful to garner local support. The attitude of the planning authority to these buildings was slowly shifting during the period 1987 to 1990, as cases 3 and 4 show. There are echoes of the golf course boom here: as a number of applications for former agricultural assets, in these cases for buildings, come into the planning department a policy is needed to guide decision-making. In the beginning the policy seemed to be structured around employment and diversification considerations and thus permissions were given, even for the conversion of pig units. As we saw in case study 1, the planners hoped that the types of economic activities that would go on in these new units would be of the traditional rural kind e.g. craft- and agriculture-related. However, it seemed to the unit managers (the ex-farmers) that light industry would be more appropriate, as this type of business would be able to afford the rents and would "blend" into the overall conception of the estates.

There can be little doubt that the development of industrial units on farms in Aylesbury Vale represents the growth of relatively new market demands for small-scale, flexible, leased workplaces as these were difficult to obtain in the nearby towns, despite there being an overall surplus of industrial warehousing space. Workshops below 1,500 sq. ft (170 sq. m), in particular, were more difficult to find on urban industrial estates, and hence they excluded small businesses working on tight and often highly leveraged margins. Moreover, the cheaper rents in rural areas allowed these new premises to undercut urban units and they provided flexible leases as well as environmentally pleasing sites for their occupiers. The industrial units themselves exhibited a variety of forms, ranging from small-scale manufacturing to storage and information technology businesses. Many of these firms were "first generation" or start-up businesses established by managerial and professional people who had moved away from their established employment to "go it alone". The prospect of "reversed commuting" and the growth in information technology (e.g. the fax), reducing the needs for an urban based location, reinforced the movement of small businesses from the neighbouring towns. While the provision of space for small businesses, as recognized in planning documents, was largely ignored as an emerging rural phenomenon, the evidence analyzed here tends to suggest that this demand is growing. Furthermore, the phenomenon is actively promoted by such bodies as the Rural Development Commission who see their rôle as catalysts in promoting small business development in the countryside, however buoyant the regional economy may be. In areas such as Aylesbury Vale they have become aware of the growth in demand over recent years and have been encouraging farmers to convert buildings. They have tended to place aesthetic landscape considerations below the encouragement of rural enterprise.

For the farmers, (i.e. the developers) once the initial investments had been made (in the provision of services, infrastructure and subdivided buildings) they simply became "landlords". They negotiated leasing and rental arrangements on a one-to-one basis. The gaining of planning permission provided the major hurdle in the re-commodification of their buildings and success placed them in the position of landed capitalists rather than petty producers. The ability to be able to define the "redundancy" of former agricultural buildings (often here assisted formally by consultants and the RDC) was a crucial stage in bringing about this transition. The case studies demonstrate that once they *disengage* from agriculture some farmers have the capability of becoming part of a new landed interest. Aided by networks of professional advice and credit finance they promote a discourse of entrpreneurialism and development. The establishment of small industrial estates on farms thus represents the emergence of a partial re-industrialization of rural areas with former farmers retaining their interests in the ownership of property but attracting combinations of high tech and low tech firms employing a variety of types of labour.

While these economic shifts presented farmers with the opportunity to adapt their buildings, the formation of a district-wide Rural (Local) Plan allowed a more strategic local planning policy to emphasize "protection" of the landscape and "quality" of the environment. This was supported in the late 1980s by government circulars which reinforced the need for local planners to adopt these objectives (DOE 1989b), and by regulations bringing new agricultural buildings under planning control. As our last case study, presented below, identifies, the *Rural Areas Local Plan* for Aylesbury Vale enshrined the shift in a restriction on industrial developments in the countryside. Strong presumptions were made on aesthetic grounds against the redevelopment of "non traditional" farm buildings. In addition, restrictions were specified where buildings were less than ten years of age or above 465 sq. m in area and where they may have detracted from the landscape of the countryside. Where redundant farm buildings were regarded as "traditional" they were still to be assessed on the basis of rigorous criteria associated with aesthetic compatibility and minimal impact. In addition, to be considered positively for planning permission, applications would need to be supported by a survey demonstrating structural integrity and financial viability. They would also have to be located outside one of the designated Special Landscape Areas. Buildings which were deemed to require substantial reconstruction (and hence capital investment) were not regarded as suitable for development, whatever their age or type.

Case study 5: Mr B., 1985–90

Mr B., in the process of disengaging from agriculture, made two applications for changes of use to his farm buildings. The two proposals span the period of change in planning policy.

Mr B. established a turkey farm in the early 1960s, producing hatching eggs (at its peak he produced half a million eggs per annum). He reared all his own breeding hens. In total, this required extensive building space of 30,000–40,000 sq. ft (3,300–4,500 sq. m). He erected two large purpose-built turkey sheds on the holding and remained in the turkey trade until the mid-1980s when the market went, in his words, "into disarray". In early 1985 he began to explore the viability of other enterprises to follow his impending move out of turkey production. Word of this spread around the locality and at about this time he was approached by various local business people looking to rent floorspace. For Mr B. this wasn't an entirely novel idea. Back in the early 1980s he had let some floorspace to a friend, in business locally, to store boxes. The arrangement had lasted for 6 months until he needed the space for turkeys. When he was approached for the second time he already had some idea of the course of action he was embarking upon.

Before making any commitments Mr B. went to see another farmer in the area who already had operating industrial units. He was told that he would have no trouble in filling his building as, effectively, demand for units outstripped supply. Furthermore, Mr B. was quite restricted in what he could do with his holding. Only having 30 ha of land, which was suitable only for grass (rented to a local grass merchant), meant that his only convertible asset was the purpose-built turkey sheds and the other assorted buildings on the farm. The conversion of these buildings into workshops and storage spaces was one of his few viable options.

The turkey sheds were relatively easy to convert. Having made the decision to go ahead, Mr B. set about putting windows in the buildings and getting three-phase electricity on site. He insulated them all and put down concrete floors in the best ones. The size of the units was partially determined by the construction of these buildings. When the pole barns had been built they consisted of bays of 420 sq. ft (45 sq. m) for the turkeys, so this formed the unit size. In the other sheds he had smaller units of 200–250 sq. ft (22–28 sq. m).

The first tenant arrived at the end of 1985 and by the end of 1986 he had filled every building apart from the large laying shed (20,000 sq. ft, 2,250 sq. m), which still housed his remaining turkeys. By this time there were 18 small businesses employing 55 people on the site. The units brought together a wide variety of firms, including high tech services, reproduction furniture manufacturers, upholstery makers, builders, frozen fish suppliers, sign-writers and garden shed suppliers. These were in general low rather than high tech firms.

Mr B. believed that the overriding factor which explained the demand for his units was rental price. He claimed his units were probably the cheapest on the market (up to 50% cheaper, he claimed), and were certainly cheaper than those on the industrial estate in the nearest village. As he used unsophisticated buildings he could charge low rents. He also eschewed formal leasing arrangements, making verbal agreements with all his tenants. They knew what the rent was going to be for three years. After that date there was a "rent review". Providing they paid their rent during that period they were left to get on with their work. At the time of final interview, in the midst of the recession, two companies had gone bankrupt and three had moved. He had not had to take on any more tenants, however, as existing businesses expanded to occupy the space.

Initially, Mr B. ignored planning. It was the visit of an enforcement officer (after the District Council was alerted by the local Parish Council) that made him formally apply for change of use. After taking advice from a neighbour who had industrial units, he employed a planning consultant to oversee the application. They decided in the first instance to apply for all his buildings except for the large turkey shed which was still vacant. He argued: "Our thinking at the time was that if we went for everything at once we might not get anything. If we went for some we might get some". They also assumed that if the first application went through a second would automatically succeed.

A formal application was then made for change of use from agriculture to warehousing, storage and industry. This went before Planning Committee in autumn 1987. Taking into account government circulars and guidance notes (e.g. DOE Circulars Industrial Development 8/87 and Development Involving Agricultural Land 16/87, and PPG 7 on Rural Enterprise and Development), as well as MAFF encouragement for farm diversification, the District Planning Officer took the view that: "From a planning standpoint current Government circular advice encourages the use of redundant farm buildings for commercial purposes, and it would appear that these buildings are answering local need for small business premises. It is recommended that the application is accepted" (AVDC Planning File). The application went through Planning Committee.

While this process was under way, Mr B. worked on his large turkey shed. He insulated the roof and built a floor covering. By mid-1987 he had moved tenants in and had filled the building. One tenant (a trailer tent distributor) took up the majority of the space – 12,000 sq. ft (1,300 sq. m) – while two more (a marquee and a marketing firm) took the rest. Six people were employed.

The planning department was aware that Mr B. had plans for this remaining building and advised him to put in an application. In early 1989 he applied for change of use to warehousing and storage (he did not risk suggesting industrial units). He claims that their initial response was favourable but at some point their opinions changed; when the proposal went to committee in late 1989 they recommended refusal. On this occasion, the Planning Officer argued that:

> Current government advice on the re-use of redundant buildings
> was not intended to include inappropriate activities in the past . . .

[moreover] . . . the council has recently adopted the policy set out in *Conversion of traditional farm buildings in open countryside* (AVDC). Structure Plan policies are against large-scale commercial developments in the countryside. It is clear that the buildings to which this application relates would not fall within the scope of that policy as they cannot be described as traditional buildings (neither would the buildings on the site which already enjoy permission for commercial use)." (AVDC Planning File)

Not only were the planning authorities indicating a marked change of emphasis in policy, they were also close to admitting that the original decision had been a wrong one. Mr B. was flabbergasted that they had turned down a building exactly the same as one for which he had previously gained permission. As he put it: "It would appear without any doubt at all that there had been a change of policy, a change of thinking between the time of the two applications".

Believing that he had a strong case, Mr B. then hired another consultant and took the case to appeal. Once again they believed they would get through for, as his consultant put it, "we felt it was virtually an open-and-shut case and so did everybody else we talked to". However, the Inspector upheld the Council's decision and turned the appeal down. The Inspector argued that the Structure Plan contained a presumption against the re-use of redundant buildings except for *small-scale* industrial activities. Furthermore, there should be some need for buildings for employment use within the immediate area. The Inspector decided that the building represented a significant increase in floorspace and could not therefore be considered small scale, while the Council had made sufficient provision for industrial premises in the area. The Inspector also echoed the Council's concern that the building's life span should not be prolonged.

Mr B. and his consultant then took legal advice to see if they could challenge the Inspector's decision, but could find no grounds to do so. Mr B. made a final attempt to salvage something from the affair and approached the planners in order to explore the possibility of a compromise – he would reduce the size of the building by half if they gave him planning permission on the rest. In Mr B.'s words: "They [the planning officers] don't want to talk to us about anything at all. So as far as I'm concerned the building *is there*, what I'm going to do with it I just don't know. Its almost as if they are saying well if we don't give it permission it'll go away . . . I mean I really can't sit here and watch it fall down . . . I don't know what I can do"

The protracted nature of this second development proposal exemplifies the changing nature of local planning policy vis-a-vis the conversion of redundant agricultural buildings. The consultant explained the shift in these terms:

When his [first] application went in it was at a time when the government were facing concern over the level of farm incomes and wanted to underwrite them, when the environment was not quite such a concern. Agricultural incomes are still, if not a greater, concern but the environment has come up so much on the political agenda that they've tended to back-pedal rather on encouraging farmers to develop in the countryside. I think it's less encouragement rather than tacit approval."

This was reflected at the District level in the *Draft Rural Areas Local Plan* (drawn up during the period of Mr B.'s applications) which placed stronger emphasis on the protection of the countryside "for its own sake". It represented a comprehensive presumption against industrial development unless the buildings were of historic and aesthetic interest. This clearly did not include modern farm buildings. The Plan stated that:

Modern farm buildings, by virtue of their size and appearance, frequently intrude upon the landscape of the countryside. They are often of utilitarian character comprising simple framed structures, clad in sheeting or blockwork, and such that substantial works amounting virtually to rebuilding would be required to make them suitable for an alternative use. Buildings, of any age or type, which can only be brought back into use by complete or substantial reconstruction are not suitable for retention; nor are those which detract from the landscape and owe their existence to an essential need which no longer exists." (para. 4.42)

However, what this case shows is that some modern farm buildings can be converted to other uses at a relatively low cost and can, therefore, fill a market niche for which other more expensively converted buildings would be too highly priced. As far as Mr B. was concerned he was supplying units at the bottom of the price range primarily because the buildings were already there and conversion costs were low. Furthermore, he claimed the cost of converting traditional buildings was much higher so the rents would therefore be higher: "That is for the people who can afford to move in and pay £5 a square foot rent. The people who come here are not that sort of people. I am dealing with a different type of person altogether" In fact he claimed the planning authorities were not concerned about this issue at all: "economics doesn't come into it"

This was borne out by examination of the *Rural Areas Local Plan*. Earlier concerns about spreading the variety of employment in the countryside were now to be balanced against "protecting the open character and the appearance of the countryside" and "the Council considers that there should be a presumption against the establishment of new such [development] uses in the countryside but existing uses should be permitted to continue as long as their impact is reduced or not made worse" (para. 7.210).

So it is primarily on aesthetic or architectural grounds that decisions about future uses of agricultural buildings are likely to be made. To quote from the *Rural Areas Local Plan* again:

> In assessing its value to, or impact on, the character of the country-side, a building's age, method of construction, and the materials used are of the greatest significance. It is considered that brick, stone and timber buildings with slate, clay tile or thatch roofs are the types most likely to be acceptable. These are buildings which most people will find attractive in the countryside, but not every building will make a sufficient contribution to the character of the countryside to justify accepting its conversion. The visible conditions of buildings, whilst obviously a factor, may not be as important as at first appears to be the case. Buildings can through lack of external maintenance become very untidy in appearance, but remain structurally in good order and may be of historical or architectural importance. (para. 4.44)

Modern farm buildings do not meet any of these criteria so it must be assumed that in the future it will be extremely hard to find alternative (non-agricultural) uses for them. However, many of these buildings now have no economic use within the sector that gave rise to them, as this case demonstrates. They are a form of residual capital, stark monuments to the age of agricultural productivism, now seemingly passed.

Again, the farmer in this case was concerned to withdraw from agriculture and as an intensive poultry farmer really only had his buildings as re-useable assets. After initially using them for storage he consulted a neighbouring farmer about industrial units and then began to convert his buildings. In this instance planning permission was only sought on the first conversion *after* the buildings had gone into their new use. Despite this contravention of planning rules he was successful. Clearly employment and diversification were uppermost in the concerns of the planners. By the time of the second application, however, things had changed. The planners could now see industrialized agricultural buildings being put to new uses and this was a trend which they believed should not be promoted. In both the district authority's decision and the appeal the forward plans, adopted at both county and district levels, provided enough ammunition to kill the proposal. This was done by stipulating certain *physical* criteria for buildings which would raise the costs of conversion and therefore restrict the amount of redevelopment likely to take place. Thus, physical considerations come to have social consequences.

The shift in planning policy documented here represented, at a broader level, the ascendance of certain aesthetic representations of the countryside over previous economic ones. The "modern" agricultural buildings were erected with little thought being given to their aesthetic impact. Now that their agricultural utility has disappeared their "intrusion" into the "aesthetic" landscape becomes even more visible. They do not conform to the new "character" of the countryside. New economic activity should not disturb the aesthetic qualities which define the value of this countryside and which will be the yardstick against which all future development will be assessed.

While the local council was not proposing to relocate the now considerable number of isolated rural employment sites scattered throughout the district, it was formally resisting the growth of new employment uses or their intensification. This was likely to concentrate pressure for industrial sites on the urban areas and to inhibit the decentralising economic tendencies in the district. There was a general presumption against any form of employment development which was proposed outside designated industrial estate land, whether this was associated with industrial, office or warehousing developments (i.e. within B1–B8 of the use classes).

The development of industrial units on farms is one significant way in which property rights, and the rules governing those rights, are tending to realign not so much along sectoral lines but increasingly across the traditional sectors of agriculture and industry. Local planning is increasingly grappling with the implications of inter-sectoral changes in property rights, and the impacts these changes imply for the establishment of a protectionist rural planning policy. Planners ran the risk of coming into conflict with the Rural Development Commission on these new rules of transfer. Both agencies seemed to be developing increasingly different conceptions of the rural economy. Planners came to be concerned with protecting the property rights and positional goods of middle-class owner-occupiers through their emphasis on aesthetic protection of the countryside "for its own sake". The Rural Development Commission adopted a more entrepreneurial standpoint, emphasizing employment and development rights within the context of environmental safeguards. The growth of private-sector consultants in rural areas (such as Aylesbury Vale) also became increasingly important for farmers wishing to gain planning permission. A high level of professionalism in negotiations was required. Thus, when farmers attempted to "go it alone" or rely on informal mechanisms for developing property (as in case study 3) they often ran foul of planning restrictions.

At the micro level, all five case studies illustrate different constellations of interests based on varying combinations of political and economic resources. The networks of actors – those supporting development (farmers, consultants, agents, and the RDA) and those questioning it (planners and local residents) – tended to marshall sets of arguments based upon particular conceptions of the countryside and the former attempted to exert political and economic pressure on planners to allow the realization of their assets through diversification activities. What was significant here was both the regulation of the development rights and the inability of planners to influence rights of occupation. Farmers could argue that if developments did not proceed, buildings would become derelict, especially as returns from agriculture were falling. Thus, the negotiations with planners were partly structured around different combinations of private property rights and the abilities and inabilities of local planners to direct development rights.

From the perspective of the farmer-developers the planning system was seen as impeding agricultural and rural diversification and not necessarily maintaining environmental protection. The planning system could, through its emphasis upon the siting, aesthetic protection and quality criteria, regulate not only industrial decentralization but also the patterns and degree of agricultural disengagement, by conferring or withholding the development rights that enabled farmers to escape the rigours of agricultural commodity markets. While planners may have considered their powers to be unduly restricted to land-use considerations, it is clear that in areas such as Aylesbury Vale they can play a key mediating rôle in influencing the shape, pace and direction of economic development in the countryside. As the case studies illustrate, their approach came to emphasize the protection of the physical landscape and traditional conceptions of the maintenance of rural life, in line with the aspirations of the residential middle class. In response to criticisms that the *Rural areas local plan* was too inflexible in its approach to industrial development in the countryside the Council replied thus: "The suggestion that the weight given to the protection of the countryside is too great with the Local Plan is not accepted" (AVDC 1992: 192). Thus, protection increasingly "wins out" over employment generation in an area so traditionally economically buoyant as Aylesbury Vale.

Conclusion

In this chapter we have stressed the rôle of development plans in shaping the uses to which rural spaces can be put. We have argued here that industrial units on farms represented, in the beginning, a regeneration of the rural economy. As the developer in case study 1 noted, in an area dominated by commuter villages, which are often almost deserted during the day, the location of many small businesses on farms represents an attempt to put both economic and social life into areas where the dominant economic activity, agriculture, had for many years been in steady decline, at least in terms of employment generated. However, for many rural residents (see Table 2.11, where 91% professed themselves opposed

217

to industry in the countryside with only 8.6% in favour), and certainly for the planning authority, these new uses of the countryside were unwelcome. The planners saw various problems with industrial estates on farms: they gave renewed life to farm buildings which were agriculturally redundant and which would not have been given planning permission in the first place, thus they began to insist that only "traditional buildings" were converted; these new estates would result in "reversed commuting" and traffic problems on country roads would increase (although it should be noted that light industry only tends to attract light vehicles while agriculture attracts large machinery and lorries); these estates do not conform to standard conceptions of the rural economy, thus the hope was expressed that the units would be used by craft- and agriculturally-related industries. It would seem likely that the concerns of the planners would be echoed by those rural residents in our survey who so opposed new industrial activities in the countryside.

The overtly aesthetic and landscape concerns now codified in the local plan can be seen as representing the views of those who see the rural as a living and recreational space. This concept of the rural outweighs any concern that it should be an economic space, unless of course the economic can be equated with a rather traditional type of agriculture or other supposedly "rural" activities. More than any other land-development process considered in this book, the case of industrial units on farms illustrates the degree to which the consumption concerns of planners and residents have blunted attempts to develop and diversify the rural economy, so evident during the mid-1980s. Thus, the plan may prove much more effective than site-specific action group activity in determining the character of the countryside in Aylesbury Vale. Within the plan certain representations of rural space, handed "down" from government and handed "up" from influential residents, are stressing that the rural must be preserved "for its own sake". In our view, the new rurality that is coming into being, which stresses aesthetic and residential values, and which seeks preservation rather than development, is best considered as an outcome of the contemporaneous process of middle-class formation. In such a traditionally prosperous area as rural Buckinghamshire this effectively *places firm limits on the extent of industrial development in the country-*side.

CHAPTER 10

A reconstituted rurality?

Whichever class is spatially (that is, locally, regionally, nationally) dominant has an interest in and pursues practices which seek to sustain and reproduce that dominance.   (Cooke 1985: 215)

Introduction

It is tempting when concluding a locality study to stand back and see what one has learned about the world in general. Having subjected one particular place to in-depth analysis, a feeling prevails that the story cannot end there. Surely issues have been raised, subjects discussed, trends discerned, that are of much wider application? In one sense the whole rationale for place-specific study is what it can tell us about other places at other times. This need is even more pressing when case study work has been conducted, for such a methodology reinforces the particularity of the account. In order to justify close scrutiny of perhaps unique events, the worlds of others, distant in space and time, must be woven into the story. We must, therefore, seek the general in the particular. In this way, the locality becomes a microcosm.

In the previous chapters we have, of course, been aware of this issue and have sought, where possible, to spell out how our Buckinghamshire findings might be more widely applicable. In this concluding chapter, however, we wish to emphasize more strongly the extent to which we believe this study to be of wider relevance. We hope to strike a fair and reasonable balance between the general and the particular but we are not seeking to argue that the "world" or even, more realistically, rural England, can be reduced to Buckinghamshire. In fact, rather than confine this "balancing act" to its spatial components, i.e. the uniqueness of the locality, we wish to shift the emphasis to *process* and ask how widespread are the patterns of social action that we have witnessed?

In order to shift attention away from place *per se*, we first return to our conceptual understanding of the most appropriate way to situate a locality in its wider context (for more detail see Marsden et al. 1993: Ch.6). We adopt a view of the locality as a "meeting place"; that is, particular spaces are seen as being "cross-cut" by a whole variety of social relations that extend in varying degrees across space and time. Thus, the locality is bound into its context in a variety of complex ways. We go on to summarize the preceding chapters and examine the variety of social processes that have cross-cut the Buckinghamshire locality. These processes have differential impacts, but they leave a series of "deposits" (to use Massey's (1984) term) which add up to a sequence of development trajectories. We have used the notion of class formation to assess the development trajectory of the case study locality. Although this is a rather broad and sweeping analytical framework, it does allow us to present our rather diverse findings and the range of likely consequences in a coherent manner. We thus conclude by considering the extent to which processes of class formation are changing more generally the nature of rural space.

The locality in context

In choosing the *district* of Aylesbury Vale in the *county* of Buckinghamshire as our area of study, we adopted a pragmatic concept of locality, one derived from political boundaries. Having thus chosen our study area we cannot assume that any other features of the area will necessarily be contained within such boundaries; there is no reason why economic, social, cultural and political processes should exhibit any kind of unity at the local level. Traditionally, the difficulties associated with defining localities have focused attention on the construction of local spaces within unfolding general economic, social and cultural processes. This usually leads on to a consideration of which particular elements of economic and social life are localized and why. The pursuit of these questions inevitably results in a disaggregation of spatial entities, for different elements of local life are differentially tied into sets of wider socio-spatial relations. In our earlier consideration of this issue (Marsden et al. 1993: 135–47) we quoted Massey (1991: 28) to the effect that:

> what gives a place its specificity is not some long internalized history but the fact that it is constructed out of a particular constellation of social relations, meeting and weaving together at a particular locus . . . It is, indeed, a meeting place. Instead then of thinking of places as areas with particular boundaries around, they can be imagined as articulated moments in networks of social relations and understandings, but where a large proportion of these relations, experiences and understandings are constructed on a far wider scale than what we happen to define for that moment as the place itself, whether that be a street, or a region or even a continent"

The challenge is thus to specify how social relations are forged and to show how actors in different localities are bound into networks of relations that determine or condition their actions. Massey terms this "power geometry" (1991: 25), referring to the way that certain actors are able to impose certain types of relations upon others. The rôle of analysis is, therefore, in the words of Callon et al. (1985: 10), to study "categories and linkages, and the way in which some are successfully imposed while others are not". These linkages may be "local" (i.e. extending only short distances), or they may be "global", (i.e. extending over long distances). We can therefore specify two dimensions of the term locality: first, it may be a given geographical space that will be cross-cut by various networks of relations; secondly, it is a local space *within* a network, often termed a "locale" (see Agnew 1993: 263). The latter understanding of the term more accurately refers to the incorporation of local *actors* within networks of relations, and so requires a sensitive handling of both the spatial and the social. In what follows we illustrate both dimensions of the term, for in this book we have sought to describe how a given geographical locality (Aylesbury Vale) has been cross-cut by various sets of relations and we have also sought to "get inside" those relations to understand how they are constructed.

It should now be clear that the relationship between a given locality and its context is necessarily complex. Either the geographical locality is cross-cut by a whole variety of different networks of relations (locality as a "meeting point") or local spaces and actors are incorporated within these networks. In the rest of this chapter we first examine Aylesbury Vale as a "cross-cut" geographical space, using examples from the preceding chapters. We then go on to examine the ways in which actors within the locality are incorporated into networks of relations, some "local", some "global".

Aylesbury Vale in context

Situated at the centre of a prosperous region, Aylesbury Vale has been subject to almost continual growth pressure over the past twenty years, this easing only with the most recent recession. This growth was caused partly by its locational advantages such as good transport links and the effects of the green belt at its southern border. Initially, in the 1960s, population growth was facilitated by the designation of London overspill towns in the area. However, the momentum has been sustained through the District's position in the "golden crescent" of economic prosperity. This subregion has become the favoured location for the new service sector and high tech industries. As these sectors have prospered, so increased economic and social pressure has come to bear at the local level. The planning system has played a crucial rôle in channelling this pressure into key centres. In Buckinghamshire the policy has been to shift growth northwards into Aylesbury and Milton Keynes. Moreover, the planning system has also sought to preserve the distinction between town and country. So although Aylesbury town has swollen, the rural areas of the district have remained largely free of any substantial population increase.

Although the existence of the planning system and Aylesbury Vale's location in the prosperous subregion largely account for patterns of development in the area, we should bear in mind Thrift's (1987b) point that development is driven not simply by economic considerations but by the social advantages to be enjoyed in particular areas. As we saw in Chapter 1, it has long been recognized that the middle-class yearning for the rural idyll has been a key force for change in rural areas. With the planning system attempting to preserve the boundary between town and country, we can expect those with the requisite resources and aspirations to seek out their place in the country. It is no accident that the "golden crescent" has desirable rural locations available to those who work in the new industries in the booming towns. Thus, as these areas begin to take on such patterns of development – professional jobs and facilities, plus middle-class rural residences and neighbourhoods – we should expect this to have a cumulative effect as those people taking locational decisions come to value living in the rural environment. Our survey data revealed that work and the environment were the two most important reasons for moving into Buckinghamshire villages, even though employment opportunities are simply not available in such villages.

In Chapter 2 we portrayed Aylesbury Vale as a locality which had benefited greatly from the general shifts in the economic and social structure of the South East. However, as our case studies have shown, these general changes have to be mediated through specific sets of relations or networks. Thus, for instance, people cannot move into rural areas unless housing becomes available. Housing becomes available through specific development–planning mechanisms, and we examined these in Chapter 3. We characterized the residential development process as a "pyramid". At the apex are the OPCS population statistics which give overall levels of "demand". These are then passed down and translated into more specific "targets" through SERPLAN regional guidance and the structure plans to local plans and housing land availability studies. In Buckinghamshire the effect of this mode of regulation, together with the more general policy of channelling growth into selected centres in the north of the county, has been to concentrate house-building around Aylesbury. This policy provides developers with the planning context in which they must frame their proposals. In our first case study we examined the provision of one new estate on the edge of Aylesbury (Watermead) which sought to provide a high level of "value-added" housing; that is, an estate where the properties and the environment were marketed as qualitatively distinct from more common types of provision. The development company specialized in "niche" production, using a range of features fused into a "lifestyle" complex of housing, environmental features and amenities. Thus, a parcel of "pseudo-Edwardian" space was constructed in a locality that was more used to homogeneous "Barrett-type" estates. Here the "rural" was reproduced in order to make the development distinct and stand apart. The estate was clearly aimed, initially at least, at a particular social stratum, characterized by an executive of the development company as the "white GTi set", by which he meant younger members of the affluent middle class. The ski slope, cricket pitch, piazza and other features were designed to attract those seeking a home that doubled as a "holiday retreat". The developer was partially successful and there was some evidence that Watermead's residents were relatively young and affluent. However, the recession slowed sales, placing a great deal of pressure on the developer, and forced the company to sell land to other house-builders. Although the latter paid some attention to the overall concept, they were building standard suburban houses. Thus, the estate was gradually losing some of its distinctiveness.

Despite high levels of house-building, with the onset of recession house prices remained relatively high. In the villages, where fewer houses were built, new housing was likely to capture even more of a premium. (As we saw, one of the reasons why Watermead was designed in the form of a new village was to capture purchasers who could not find an "authentic" village residence). The second case study examined the conversion of barns to housing and found that these were selling for prices out of the reach of all but affluent incomers. These houses appealed to a rather older strata than those of Watermead and they allowed the "heart" of the village to be reconstituted in more traditional fashion. By sympathetically converting existing buildings, some continuity of form was encapsulated in a physical reference to a rural past. The purchasers of these conversions also seemed to value moving into an existing (rural?) community, where a network of relationships could be obtained along with the house.

The problems of access for those without the resources to compete in the housing market were illustrated in our final case study. In Weston Turville a few young residents, often born in the village, sought low-cost housing. The only way this was likely to be provided was as an "add-on", or more cynically, as a "sweetener" to a larger development. However, such a proposal was unacceptable to the majority (or at least the most vocal) of residents. This case illustrates the sheer difficulty of providing low-cost or rented housing and how social selectivity is imposed by the residential development process.

While housing developments allow some new residents to enter rural areas, such developments must take their place in an existing social and physical environment. In Chapter 4 we examined three contrasting village environments. We looked first at Swanbourne, a village in the heart of the district. Here the estate landlord still held sway and several small tenant farmers dominated the village structure. The amount of external development in Swanbourne was limited, and the landlord, in conjunction with the planning authority, seemed determined to sanction only a gradual evolution in Swanbourne's social and physical structure. Thus, new entrants, of which there were few, found themselves in a traditional rural community where deference and continuity were still evident. This village represents a form of rurality which might be described as "traditional": a picturesque place dominated by agriculture but where the sense of community was subdued, as if the dominance of the landlord had allowed the majority of the residents to drift into apathy. There were insufficient in-migrants to significantly alter the prevailing ambience: thus the timelessness and inertia of Swanbourne looked set to continue.

We then turned to Weston Turville, a "suburban" village in the south of the District. Here high levels of development pressure and house-building had been experienced and the village had grown rapidly throughout the 1970s and 1980s. The consequence was a "village" of many small communities, orientated around particular village institutions, held together mainly by women. But given the numbers now living in the village, there were clearly many residents who lived a private suburban life. Although Weston Turville clearly exhibited the qualities of middle-class colonization, the reconstitution of the rural qualities of the village were almost swept away by the relentless processes of development. There was still a feeling in the village that it should hang on to what vestiges of rurality remained, as the glebe land action group demonstrated. However, the village seemed to be teetering on the edge of suburban anonymity. Where Watermead represents the ruralization of suburbia, Weston Turville exemplifies the suburbanization of the rural.

Our last village study was Wingrave. It is here that we see the reconstitution of rurality most clearly. Wingrave retained many traditional features, particularly within the village "heart" and was still home to many "traditional" residents. However, the establishment of a new estate, and the withdrawal of farming from the village boundaries, had opened up a space for an appreciable number of in-migrants to settle. These fell into two middle-class strata: middle-income families and those in the higher income groups. The former tended to move onto the new estate, whereas the latter settled in the older part of the village where most of the traditional buildings were located. With the wave of incomers came a revitalization of village life. The number of local organizations and societies blossomed and a sense of community began to emerge. But this community was forged by the more affluent of the incomers and thus the identity of Wingrave came to be closely associated with a particular type of class identity.

The portrait of Wingrave that we have presented here bears more than a passing resemblance to Pahl's findings in Hertfordshire during the 1960s. As in Pahl's study, the process of middle-class in-migration had brought different classes into close proximity with one another in the village. The marginalization of some groups such as the council house tenants was perhaps more advanced than in Pahl's case, but all were still aware of the presence of the others. However, there seemed to be little contact between the different classes, as the interviews with the council tenants illustrated. It is likely that as development slows (and with the planners seemingly determined to keep the village in its present shape) the level of house prices will remain high and over time the hold of the middle class on the village will be strengthened. Thus, Wingrave, more than either Swanbourne, which still retains a parochial and traditional social order, or Weston Turville, which has been swamped by development and incomers, provides fertile ground for a middle-class community to come into existence; that is, a middle-class space that allows a rich array of cultural forms to solidify middle-class identity.

We can begin to see here how forms of economic and cultural capital come to be invested in property and cultural assets and how these become intertwined in particular places. The village is the site for property investment, but what makes it attractive is not simply the state of the houses but the cultural environment in which residents will live. The culture of the village, as exemplified in Wingrave, allows a sense of middle-class unity to be achieved which may be more difficult to attain in urban and suburban locations. The processes of "distinction", whereby members of one class can "mark" themselves off from members of other classes, are easy to see in the rural context. First, the village itself is physically distinct from other settlements, allowing assessments of its social status (positionality) to be readily made. Secondly, although within a village members of different classes may be in close proximity to one another, as Wingrave shows, patterns of residential segregation are still clearly observable and there need be little interaction between residents of the different areas.

The village thus allows a variety of opportunities for interaction between new middle-class arrivals and this can result in activities that exacerbate the class character of these spaces. We might then expect to see the emergence of forms of (political) action orientated towards the protection and enhancement of middle-class cultural and property assets. In Chapters 6–9 we considered how the crisis in agriculture was allowing and directing such actions to determine patterns of development in the wider countryside. Thus, the villagers begin to impose their cultural aspirations beyond the boundaries of the village. Consideration of middle-class actions around specific development proposals allows us to show how the social and spatial are asserted within particular development networks. Thus, we now turn to examine how the locality is the outcome of actions within these networks of relations.

A reconstituted rurality?

Making the "local" and the "rural":
development and anti-development networks

We use the term "network" here to describe how groups of actors come to be aligned over specific issues, in this case development projects. That is, sets of actors join forces either to promote or to resist development. They can therefore be considered as bound together in a network. In so becoming bound in they promote certain conceptions of the "rural" and the "local": they seek to represent particular spaces in particular ways. If the development network is successful, it will bring into being a new social space (such as Watermead); if the anti-development network is successful (as in Weston Turville over the protection of the glebe land), it will preserve a particular social space. The outcomes deriving from network activities will begin to "stack up" in given localities; thus, the locality begins to take on a certain social and spatial shape and this provides the context for new "rounds" of network activity. We can illustrate this process by reference to the cases presented in Chapters 5–9.

In Chapter 5 we saw how farmers were seeking to diversify their business activities in order either to leave farming altogether or to stave off financial problems. Farmers were beginning to disengage from many of their traditional networks – food supply industries, agro-chemical suppliers, abbatoirs, supermarkets, etc. – and were drawing upon external forms of advice – land agents, ADAS, bank managers, etc. – in order to find new market arenas in which to participate. This marks a shift in the complex web of relations surrounding rural land use. However, it also gives the planning system a crucial rôle to play in determining how much land can leave agricultural production.

The most successful new use for rural land in Aylesbury Vale, at least in terms of acquiring planning permission, was for golf courses. A rather bland explanation for this phenomenon might be along the lines that "with so many middle-class inmigrants the demand for golf was high so new courses were needed". However, as our case studies showed, the situation is more complex than this. During the late 1980s a diverse set of considerations were woven together to provide a prevailing common theme of "unmet demand" in the golfing sector. Concerns over surplus agricultural land and reports emanating from golfing, development and central government organizations, all pointed towards the need for new golf courses in the countryside. As we saw, planners in Aylesbury Vale found themselves swept along on a wave of golf course proposals, supposedly bearing out the wider claims being made by sporting and development actors. Thus, local developers were able to draw upon an already existing common theme surrounding the supposed "demand" for courses. In the first case study, a farmer wishing to shift land out of agriculture began to weave together a development network from local resources. As a district councillor he was aware of local demand calculations, the planning authority's policy towards golf course developments, and the correct way to frame his proposal. He also utilized his extensive social networks to build a partnership with two brothers who lived locally. The latter put in the capital, the farmer put in land, buildings and farm labour. Thus, a driving range and golf course came into being, imposing a new set of uses and values on previously agricultural land. In the second case a rather "longer" network was established. The farmer put in the land and farm labour to the project, but to gain access to other resources he entered into a relationship with a development company. The farmer thus ceded effective control of the development. Again a new consumption space was established, appealing to affluent consumers of golf.

These two examples show two contrasting golf-course development networks and the ways in which these were able to establish new developments in the countryside. Although there was some (limited) opposition to these proposals, the planning system was (at the time) well disposed to this type of land use. Thus, both went through relatively easily. This was in marked contrast to the minerals and landfill development cases. In the former we saw how a well defined and stable minerals planning network had come into existence, binding the locality into a clear and strong discourse of demand. As in the case of housing regulation, at the apex of the minerals planning network are national calculations of future demand levels. These then pass through the various layers of the network to the county where they become quotas. Thus, when developers such as Steetley wish to open a quarry, they are able to draw upon this discourse in order to justify the disruption to local residents. In the Chackmore case we showed how the anti-development actors had to build an alternative network, again extending outwards from the locality, in order to draw in national calculations, this time associated with a national "treasure", the landscape gardens at Stowe School. The outcome was perhaps predictable once the anti-development network had been extended all the way to the National Trust and the DoE. However, if national considerations had not been incorporated into the anti-development network, those existing in the pro-development network would surely have prevailed.

In the case of landfill, a set of regional considerations were available to the development company. However, there was no stable network in existence which allowed such calculations to be transmitted unambiguously to all parties. Thus, even though the anti-development network was essentially localized, it was able to draw upon a rich social context (in terms of expertise), and could use local representations of place to defeat the proposal.

These two cases exemplify how the changing social context can lead to new forms of social action around (traditional) forms of rural development. By progressively excluding such developments, the locality comes to assume an increasingly middle-class shape. As this process accelerates, so other aspects of local life become imbued with middle-class connotations. In Chapter 9 we saw how the planning framework came to adopt a much more preservationist stance with reference to industrial units in farm buildings. Thus, the transformation of agricultural assets became more difficult as planners progressively attempted to preserve the more aesthetically "pleasing" elements of rural space by permitting new uses only for "traditional" buildings. Our survey indicated that there was a great deal of local support for this tough planning stance. Thus, a rather "traditional" form of rurality was becoming codified in development plans, one which constrained farmers in their search for new sources of income and restricted the provision of new employment opportunities.

Each of these cases shows "local" actors bound into "external" networks in differing ways. In the case of golf, planners were not formally bound into "long" networks, although they were clearly aware of a powerful discourse concerning "unmet demand" in the golfing sector emanating from more informal relationships with external agencies. The developers were also more localized, although the second case study involved a company based in London. In the landfill study, the development network was regionalized and the anti-development network was local, whereas in the minerals study the networks extended from the local to the national. Over time the outcomes of these network activities yield the shape of the "local" and the "rural". As certain representations increasingly "win out" over others, the locality comes to take on a coherent shape and this provides a structure of enablement and constraint for future action.

We have characterized this shape as "middle class" and have sought to show how the class character of the place derives from, and goes on to structure, a multitude of different situations. The activities of housing networks supply a product that takes on a distinctive class character; this allows middle-class residents to invest more and more in property and cultural assets; once the investment has been made, these residents will come together to protect their assets by campaigning against forms of development which threaten the social and natural environment in which they live. The experience of engaging in such campaigning activities enhances the sense of solidarity between residents and allows the emergence of a coherent communal identity. Such political action reinforces the class complexion of the place, for by excluding undesirable forms of development (and undesirable "others") it proves more and more attractive to those with the resources to afford property in the area. As one place becomes ever more stridently protected, so others are likely to be over-developed and thus devalued. The positional status of the former will be increased. Thus, the notion of middle-class formation allows us to discern coherence in the wide range of economic, social and political activities undertaken in an area such as Aylesbury Vale.

What type of rurality?

In the account provided here we have tried to show the complex processes involved in the making of a rural place. We have seen a whole variety of different actors – developers, planners, middle-class residents, farmers, ecologists, hydrologists, etc. – weaving together a whole variety of different elements – land, buildings, quarries, communities, meadows, lakes, etc. – in order to bring their conceptions of place into being. The result is a delightful stretch of countryside in the midst of an urban region. With Milton Keynes to the north, green belt and London to the south, the struggle to hold onto the rural in Aylesbury Vale is by no means easy. But the social make-up of the place means that a formidable array of actors can usually be assembled to orchestrate opposition to unwelcome development. As its positional status grows, the area will become even more attractive to those would-be residents who are trapped on the "outside". Thus, competition for resources, notably housing, will continue to increase, making it more and more difficult for those on low incomes to either stay in, or move to, such areas. The middle-class complexion of the locality is thus assured – at least in the short term – especially as political action and planning policy are likely to reinforce and reflect the prevailing social composition.

We have also been able to gain some (limited) understanding of what makes this place so attractive to well heeled incomers. They are looking for rural life, which means life in a "community". If such a community does not exist, it will be created, as incomers weave together the "old" and the "new" into a "hybrid" rurality, one that seeks to exclude all that these residents have moved away from, i.e. the pernicious effects of urbanism, with its "fragmented" ways of life, its "mixed-up" classes and ethnicities, its "ambivalent" sexualities. In the rural communities of Aylesbury Vale old certainties can be renewed and security re-established. Thus, the new rural communities can be seen as sites for "anchoring" traditional middle-class identities. But this can be achieved only by making such places "exclusive", by keeping out forms of development and forms of life that do not "fit" in to the new rural spaces. The more successful the processes of exclusion, the more mature the process of middle-class formation.

Although this process is undoubtedly well advanced in Aylesbury Vale, it is by no means unique to this locality. Such a general shift in the composition of rural areas is occurring almost everywhere in England, particularly in those areas within commuting distance of major centres. The attraction of the rural to the middle class has been well documented and we have tried to show here some of the likely consequences that flow from middle-class colonization. Certain forms of rural development, such as up-market housing, will be assured by the presence of large sections of the middle class, while other forms, e.g. landfill, will be progressively excluded.

It is interesting to note, however, the emergence of certain contradictions within the process of class formation. For instance, we have shown in some detail how the buoyant regional economic context, orientated towards service industries, has led to increased development pressure in the rural areas of Aylesbury Vale as members of the middle class seek out the "rural idyll". Many of these residents are employed by day in firms located in urban areas, whose accumulation ethic is the same as that which drives the development process. Yet by night and at weekends, these middle-class workers become preservationist and they wish to keep the grubby world of commerce and industry at bay. (Clearly, this attitude varies according to the type of development, i.e. minerals and landfill are likely to be far more unpopular than high tech industrial units, but using the countryside for employment generation remains contentious). Thus, the middle class straddles the divide between production and consumption; residentially, people wish to be able to shape their own consumption spaces, and these spaces extend farther and farther from their own homes into the green pastures of the countryside. Although the two poles of this contradiction were conveniently separated during most of the post-war period by a planning and agricultural regime that protected agricultural land at almost any cost, in the post-productivist period these contradictions are becoming starkly apparent. Thus, the planning system is now again called upon to protect the rural; in this instance the call is being made not for agricultural reasons but in order to placate those who have recently invested so much of their economic and cultural capital in the countryside.

However, we should remind ourselves that local planning policies are more than a simple reflection of the local processes of middle-class formation. They should also be seen "in terms of internal politicking, both between local interest groups within the state and also between central and local government" (Savage & Warde 1993: 170). We have examined the ways in which a variety of interests compete within the particular development sectors in the context of central government guidelines and policies. Much less attention has been paid here to the relationship between local and central government in the formulation of general planning policies. However, as Chapter 9 showed, the drift towards deregulation initiated in the early 1980s had by the end of the decade run out of steam, and the rural planning system could return to its traditional concerns of urban containment and countryside protection. That this is the outcome of the struggles surrounding planning in the 1980s is illustrated by the recent warning from the Secretary of State for the Environment, John Gummer, that "the Government looks to local planning authorities to breathe fresh life into the countryside through their development plans and planning decisions. Conservation alone is not enough. New jobs and wealth are needed in rural areas just as much as in towns. Without them the countryside will wither away physically, economically, socially" (DoE 1993). If the analysis presented in these pages is correct, the Government is here again attempting to swim against the preservationist tide, and this initiative, like those in the mid-1980s, will founder against the rock of middle-class political antagonism.

The context in which the processes of countryside change are being played out – a planning system with a preservationist bias, an agricultural sector in crisis, a burgeoning middle class and a nation with an Arcadian view of its past – is common to many localities across England. We may thus expect to see rurality reconstituted as part and parcel of middle-class formation in many areas. In fact, we would claim that it is this particular reconstitution which is currently dominant in rural Britain. Some areas clearly provide a fertile seed-bed in which middle-class formation can take place, while others may be much more antagonistic. However, it seems to us that the middle class will gradually exert its hold over ever wider stretches of rural land, replacing an agricultural veto with a preservationist one. This is not to say that Aylesbury Vale provides the model of development for the rest of rural Britain. This is simply one locality, where the processes of middle-class formation and rural reconstitution are well advanced, which provides some clues to the likely shape of other rural localities experiencing similar pressures.

Conclusion

We have sought to show in this volume that the processes of rural change and class formation are inextricably bound together – society and space are unified through *action*. We have chosen to illustrate this through studies of the land-development process. These studies have allowed us to examine particular instances of social action around development projects and have thus provided a detailed account of how the general becomes the specific, or how a single locality is the outcome of many complex processes. The coherence of these outcomes has been characterized in terms of class formation. We have tried to avoid using a national class category as means of explaining the variety of social processes documented in the preceding pages. Rather, we see the national class structure as an outcome of the processes of formation, some of which we have documented here. As Savage et al. (1992) remind us, class formation does not take place on the head of a pin; it takes place in *specific places* as actors come together, using the assets at their disposal in order to impose their conceptions of space upon others. The rural thereby becomes an expression of power, of the way sets of relations are drawn together and used to impose a whole variety of goals. But as these begin to converge, as they begin to reinforce one another, so some semblance of coherence is apparent. For us, that coherence can best be understood in terms of middle-class formation.

The processes of segregation and differentiation illustrated in these pages are by no means new. For instance, urban sociologists have long been interested in the processes underlying phenomena such as suburbanization and gentrification. The former, characterized as the movement of population out of the inner city, now clearly extends to the countryside, as our studies of Watermead and Weston Turville illustrate. Work on gentrification has examined the movement of the middle class into areas traditionally "home" to other (working) classes within the city. But again we have witnessed this general phenomenon in villages such as Wingrave. The aspirations that lie behind these moves to the countryside have only been hinted at in our study. We noted the desire on the part of some residents to bring up their children in an all-white, middle-class environment (Ch. 4) and saw that the "peace and quiet" of the countryside was strongly attractive (Ch. 2). The rural domain is reassuring to the middle class. It is a place where gender and ethnic identities can be anchored in "traditional" ways, far (but not far enough?) from the fragmented, "mixed-up" city. Within the rural domain identities are fixed, making it a white, English, family-orientated, middle-class space; a space, moreover, that is imbued with its own mythical history, which selects and deploys particular, nativistic notions of what it is to belong to the national culture. That this is what attracts middle-class in-migrants to the countryside is rarely made explicit. Instead, the rural is extolled for the virtues of peace and quiet, of community and neighbourliness, virtues deemed to be absent from the urban realm. We have argued that the assertion of this form of rurality necessitates the exclusion of other social groups, usually under the guise of excluding development, who as members of other – urban – cultures might legitimately make some claim on the countryside. Such groups are increasingly unwelcome in the reconstituted rural spaces of southern England.

In this book we have concentrated upon the expression of these aspects of middle-class formation within various development processes. We have examined struggles around development rather than more straightforwardly "social" issues. This is partly because we believe that it is often the way in rural England that concern for the rural environment can be translated into the desire to protect a particular social space for the benefit of a privileged social group. Fighting to maintain the rural environment, and struggling against development, is a more acceptable endeavour than seeking to exclude the less well off. That the latter is an outcome of the former is, we are asked to believe, an "unfortunate" consequence which simply cannot be helped. We have tried, in this book, to place this "unfortunate consequence" in the wider context of middle-class forma-tion in southern England. It should now, we hope, be clear that such an outcome is not unintended or accidental, but follows relentlessly from the reconstitution of rural space by an increasingly dominant class.

APPENDIX

Methodological considerations

Case study methodology

The principal methodology adopted in the analysis presented here is that of the case study. For each case study of land-development (housing: infill, barn conversions, new settlements – Ch. 3; golf course developments – Ch. 6; mineral development – Ch. 7; landfill – Ch. 8; industrial developments – Ch. 9) a series of in-depth interviews were conducted with the constellation of key participants involved in development. This involved interviews with both "locally based" actors and those operating "at a distance" ("strategic actors"), such as representatives of firms, developers, exchange agents and policy-makers. This followed the methodology outlined in Grove-Hills et al. (1990) and Marsden et al. (1993). Some of the development processes – such as minerals, landfill and golf – were unfolding at the time of the study. For others, such those associated with housing and industrial units, we retrospectively gained access to the way the participants saw the process unfolding from their particular vantage points. We have presented these case studies as accurately as possible, based on the evidence we received from both a wide range of informants and written texts such as policy documents, planning files and newspapers. We have attempted to weld this qualitative evidence together, not so much as a representative picture of land-use change in the area (clearly, there are fewer mineral sites than farms) but as contextualized examples of the key processes currently shaping new rural spaces.

Cluster methodology

Although the focus of our case studies was particular land developments occurring either before or during the period of study, we also wished to obtain a picture of how these processes "added up" in particular villages. In order to do this, our second in-depth survey strategy involved the study of three particular villages: Weston Turville, Wingrave and Swanbourne (see Fig. I and Ch. 4). For these villages we:
- conducted a series of key interviews with village "elites" (local councillors, organizers of main village activities, developers, school teachers)
- attempted a full survey of all farmers in the parish
- examined the planning and development histories of the villages through the examination of planning files and interviews, and
- conducted a small survey of household residents.

Appendix: Methodological considerations

Use of extensive and intensive surveys

Household survey

In order to contextualize this qualitative evidence, and to supplement our attempts to assess not only the features but also the social impacts of land development processes, the authors conducted (on behalf of Buckinghamshire County Council) an extensive survey of 2,000 households in January to March 1991. The households were randomly selected from 20 differing urban and rural settlements across the county as a whole (see Fig. I). The survey aimed to focus upon:

- the factors that people felt important to their living environments
- an examination of the realities of urban and rural life in the county and how these measure up to expectations and aspirations
- those factors which influence where people live and why, and
- the impact of current planning policies and the assessment of some key options concerning future structure planning priorities.

See Marsden et al. (1991) for a further description of this work.

Of particular significance for our analysis were surveys conducted in seven villages. These included:

- North Crawley: a village close to the new city of Milton Keynes with a population of just over 800; high proportions of owner-occupied housing and those in socio-economic groups 1 and 2 (49%). Development was generally restricted but it may be subject to large-scale development pressure (sampling fraction 11.3%).
- Newton Longville: a large village in Aylesbury Vale District with a population of 2,000. A former brick-making village, it has a mixed socio-economic structure with some local authority housing, and potential for development according to the structure plan (para. 12 of structure plan; sampling fraction 4.3%).
- Steeple Claydon: a village with a population of 1,800 in Aylesbury Vale that experienced considerable new development during the 1970s and 1980s. (Again a "para. 12" village; sampling fraction 5.7%).
- Haddenham: a large village in Aylesbury Vale (population 5,000); higher proportion of socio-economic groups 1 and 2 (49.3%) with high level of car ownership. Another "para. 12" village which may be capable of absorbing more housing growth, although the preservationist lobby in the village is strong (sampling fraction 1.9%).
- Lacey Green: a small village in the green belt of Wycombe district. Housing developments are restricted to the replacement of existing dwellings on a one-to-one basis, or on a very restricted infilling basis (para. 23 of the structure plan; sampling fraction 5.8%).
- Stokenchurch: a large south Bucks village (population 4,419) in Wycombe district, lying beyond the outer boundary of the green belt, with some extensive infilling being permitted. Extensions to village developments are not permitted. Considerable debate has ensued around the boundary issues concerning green belt notation (para. 26; sampling fraction 2.1%).
- Botley and Leyhill: small village (population 939) in Chiltern District and green belt. High proportions in socio-economic categories 1 and 2, (para. 23 village; sampling fraction 10.4%).

236

Farmer survey

In Chapter 5 particular use is also made of some of the extensive and longitudinal evidence collected under the EC-funded "Farm Household Adjustment in Western Europe" (1987–91) project administered by the Arkelton Trust (see the Arkelton Trust 1992) and for a more detailed agricultural survey of the Buckinghamshire survey area (Marsden et al. 1991). The surveys, overall, involved a baseline and final survey of an original 297 farmers in the county randomly selected from size class bands (see Arkelton Trust 1987). In addition over 60 farmers were tracked over the study period and re-interviewed on three further occasions (the "panels"). In these more in-depth interviews we explored land-development strategies, financial and credit links and policy uptake and assessment. In Chapter 5 some of the farm survey data is presented to show the restructuring of agricultural resources more generally in Buckinghamshire and more particularly in Aylesbury Vale.

Acknowledgements

We wish to acknowledge the support of Buckinghamshire County Council, the Environmental Services Unit, South Bank University and the European Commission, in addition to the Economic and Social Research Council in conducting this research.

Bibliography

Abercrombie, P. 1933. *Town and country planning*. London: Butterworth.
— 1944. *Greater London Plan*. London: HMSO.
Abercrombie, N. & J. Urry 1983. *Capital, labour and the middle classes*. London: Allen & Unwin.
ACORA 1990. *Faith in the countryside*. Worthing & London: Churchman & ACORA.
Adams, J. 1991. *Determined to dig: the role of aggregates forecasting in national minerals guidance*. London: CPRE.
Agnew, J. 1993. Representing space: space, scale and culture in social science. In *Place/culture/representation*, J. Duncan & D. Ley (eds), 251–71. London: Routledge.
Ambrose, P. 1986. *Whatever happened to planning*. London: Methuen.
Arkelton Trust 1987. *Rural change in Europe*. Proceedings of the first annual review meeting. Nethy Bridge: Arkelton Trust.
Aylesbury Vale District Council 1988a. *Aylesbury Vale leisure strategy: a report by L&R Leisure Group*. Aylesbury: AVDC.
— 1988b. *Coldharbour Farm: planning brief*. Aylesbury: AVDC.
— 1989a. *Aylesbury town map*. Aylesbury: AVDC.
— 1989b. *Swanbourne conservation area*. Aylesbury: AVDC.
— 1991a. *Aylesbury Vale rural areas local plan: consultation draft*. Aylesbury: AVDC.
— 1991b. *Weston Turville conservation area*. Aylesbury: AVDC.
— 1992. *Aylesbury Vale (Rural Areas) Local Plan Consultation Draft, Report on Public Consultation, Representations and Council's Response and Alterations*. Aylesbury: AVDC.

Ball, M. 1986. The built environment and the urban question. *Environment and Planning* D **4**, 447–64.
Barlow, J. 1988. The politics of land into the 1990s: landowners, developers and farmers. *Policy and Politics* **16**(2), 111–21.
— & M. Savage 1986. The politics of growth: cleavage and conflict in a Tory heartland. *Capital and Class* **31**, 156–81.
Barnes, B. 1990. Status groups and collective action. *Sociology* **26**(2), 259–70.
Blowers, A. 1980. *The limits to power: the politics of local planning policy*. Oxford: Pergamon.
Boucher, S. 1993. New settlements and community benefits: public and private responsibilities. *Journal of Rural Studies* **9**(3), 257–66.
Bourdieu, P. 1984. *Distinction: a social critique of the judgement of taste*. London: Routledge.
Bradley, T. & P. Lowe 1984. Locality, rurality and social theory. In *Locality and rurality: economy and society in rural regions*, T. Bradley & P. Lowe (eds), 1–23. Norwich: Geo Books.
Brindley, T., Y. Rydin, G. Stoker 1989. *Remaking planning: the politics of urban change in the Thatcher years*. London: Unwin Hyman.
Britton, D. (ed.) 1990. *Agriculture in Britain: changing pressures and policies*. Wallingford, England: CAB International.
Buller, H. 1984. *Amenity societies and landscape conservation*. PhD thesis, King's College, University of London.

Bibliography

—& P. Lowe 1990. The historical and cultural contexts. See Lowe & Bodiguel (1990), 3–20.

Buckinghamshire County Council 1974. *Gravel working in south Bucks*. Aylesbury: Bucks CC.

—1978. *Minerals subject plan: draft for consultation*. Aylesbury: Bucks CC.

—1980. *Approved Buckinghamshire Structure Plan: written statement*. Aylesbury: Bucks CC.

—1986. *Approved Buckinghamshire Structure Plan: written statement*. Aylesbury: Bucks CC.

—1989a. *Countryside recreation*. Aylesbury: Bucks CC.

—1989b. *Replacement minerals local plan: draft for public consultation*. Aylesbury: Bucks CC.

—1990. *Replacement minerals local plan*. Aylesbury: Bucks CC.

—1991a. *Local population survey*. Aylesbury: Bucks CC Planning Department.

—1991b. *Labour supply in Buckinghamshire to 2001*. Aylesbury: Bucks CC.

Callon, M., J. Law, A. Rip 1985. *Texts and their powers: mapping the dynamics of science and technology*. London: Macmillan.

Champion, T. & A. Green 1988. *Local prosperity and the North–South divide: a report on winners and losers in 1980s Britain*. Mimeo., University of Warwick.

—& A. Townsend 1990. *Contemporary Britain: a geographical perspective*. London: Edward Arnold.

—& A. Green 1992. Local economic performance in Britain during the late 1980s: the results of the third booming towns survey. *Environment and Planning* A **24**(2), 243–72.

Cherry, G. 1986. Settlement planning and the regional city. In *Regional cities in the UK 1890–1980*, G. Gordon Cherry (ed.), 14–29. London: Harper & Row.

Cloke, P. 1983. *An introduction to rural settlement planning*. London: Methuen.

—& N. Thrift 1987. Intra-class conflict in rural areas. *Journal of Rural Studies* **3**, 71–6.

—1989. Rural geography and political economy. In *New models in geography*, vol. 1, R. Peet & N. Thrift (eds), 164–97. London: Unwin Hyman.

—& J. Little 1990. *The rural state? Limits to planning in rural society*. Oxford: Oxford University Press.

—& M. Moseley 1990. Rural geography in Britain. See Lowe & Bodiguel (1990), 117–35.

—& N. Thrift 1990. Class and change in rural Britain. See Marsden et al. (1990), 165–81.

—M. Phillips, R. Rankin 1989. Middle-class housing choice: channels of entry into Gower, South Wales. In *People in the countryside: studies of social change in rural Britain*, T. Champion & C. Watkins (eds), 38–51. London: Paul Chapman.

Commins, P. 1990. Restructuring agriculture in advanced societies: transformation, crisis and responses. See Marsden et al. (1990), 45–76.

Cooke, P. 1985. Class practices as regional markers: a contribution to labour geography. In *Social relations and spatial structures*, D. Gregory & J. Urry (eds), 213–41. London: Macmillan.

Country Landowners Association 1988. *New golf courses: economic and marketing opportunities*. London: Country Landowners Association.

Countryside Commission 1987. *Recreation 2000: policies for enjoying the countryside*. Cheltenham: Countryside Commission.

Cotgrove, S. & A. Duff 1980. Environmentalism, middle-class radicalism and politics. *Sociological Review* **28**(2), 333–51.

Cresswell, P. 1974. *A new policy for north Buckinghamshire: submission on matters proposed to be included in the County Structure Plan*. Unpublished paper, Aylesbury County Library.

Bibliography

Crompton, R. 1993. *Class and stratification: an introduction to current debates*. Cambridge: Polity Press.

Department of the Environment 1980. *Land for private housebuilding: Circular 9/80*. London: DoE.
—1984. *Land for housing: Planning Policy Guidance Note 3*. London: HMSO.
—1986. *South East regional strategic guidance: letter from the Secretary of State for the Environment to Chairman of SERPLAN, annex A. Planning Policy Guidance Note 9*. London: HMSO.
—*Development involving agricultural land*. London: DoE.
—1988. *Assessment of environmental effects regulations: Circular 15/88*. London: DoE.
—1989a. *Guidelines for aggregates provision in England and Wales: Minerals Policy Guidance Note 6*. London: HMSO.
—1989b. *Permitted use rights in the countryside*. London: DoE.
—1991a. *Housing land availability*. London: HMSO.
—1991b. *Guidelines for aggregates provision in England and Wales: review of Minerals Policy Guidance Note 6*. London: HMSO.
—1992a. *The relationship between house prices and land supply*. London: HMSO.
—1992b. *Countryside and the rural economy: Planning Policy Guidance Note 7*. London: HMSO.
—1992c. *Land for housing: Planning Policy Guidance Note 3*. London: DoE.
—1993. Breathing life into our countryside [news release]. London: DoE.

Eckersley, R. 1989. Green politics and the new class: selfishness or virtue? *Political Studies* **37**, 205–23.
Eder, K. 1993. *The new politics of class: social movements and cultural dynamics in advanced societies*. London: Sage.
Elson, M. 1989. *Recreation and community provision in areas of new private housing*. London: The Housing Research Foundation.
Evans, A. 1991. Rabbit hutches on postage stamps: planning, development and political economy. *Urban Studies* **28**(6), 853–70.
Everton. A. & D. Hughes 1987. Minerals subject plans in action. *Journal of Planning Law* (March), 174–84.

Fielding, T. 1989. Inter-regional migration and social change: a study of south east England based upon data from the longitudinal study. *Institute of British Geographers, Transactions* **14**, 24–36.
—1992. Migration and social mobility: south east England as an escalator region. *Regional Studies* **26**(1), 1–15.
—& M. Savage 1987. *Social mobility and the changing class composition of south east England*. Working Paper in Urban and Regional Studies, University of Sussex .
Fishman, R. 1987. *Bourgeois utopias: the rise and fall of suburbia*. New York: Basic Books.
Fisherman, R. 1991. The garden city tradition in the post-suburban age. *Built Environment* **17**(3/4), 232–41.
Forrest, R. 1987. Spatial mobility, tenure mobility, and emerging social divisions in the UK housing market. *Environment and Planning* A **19**, 1611–30.
—& A. Murie 1990. Housing markets, labour markets and housing histories. In *Housing and labour markets*, J. Allen & C. Hamnett (eds), 42–64. London: Unwin Hyman.
Friends of Aylesbury Vale 1986. *Chronicle of the Vale: a "doomsday" survey of life in the villages of the Vale of Aylesbury in the mid-1980s*. Aylesbury: The Friends of Aylesbury Vale.
Galbraith, K. 1992. *The culture of contentment*. London: Sinclair Stevenson.

241

Bibliography

Gasson, R., G. Crow, A. Errington, J. Hutson, T. Marsden, M. Winter 1988. The farm as a family business. *Journal of Agricultural Economics* **39**, 1–41.

Goldthorpe, J., C. Llewellyn, C. Payne 1980. *Social mobility and the class structure in modern Britain*. Oxford: Oxford University Press.

— & G. Marshall 1992. The promising future of class analysis. *Sociology* **26**, 381–400.

— & C. Payne 1986. On the class mobility of women: results from different approaches to the analysis of recent British data. *Sociology* **20**(4), 531–56.

Goodman, D., B. Sorj, J. Wilkinson 1987. *From farming to biotechnology*. Oxford: Basil Blackwell.

Grant, W. 1977. *Independent local politics in England and Wales*. Farnborough: Teakfield.

Grove-Hills, J., R. Munton, J. Murdoch 1990. *The rural land development process: evolving a methodology*. Working Paper 8, ESRC Countryside Change Initiative, Department of Agricultural Economics, University of Newcastle upon Tyne.

Hall, P. 1984. *World cities*, 3rd edn. London: Weidenfield & Nicholson.

— H. Gracey, R. Drewett, R. Thomas 1973. *The containment of urban England*, vol. II. London: Allen & Unwin.

— M. Breheny, R. McQuaid, D. Hart 1987. *Western Sunrise: the genesis and growth of Britain's major high tech corridor*. London: Allen & Unwin.

Hamnett, C. 1986. The changing socio-economic structure of London and the South East 1961–81. *Regional Studies* **20**(5), 391–406.

— 1987a. A tale of two cities: socio-tenurial polarisation in London and the South East, 1966–81. *Environment and Planning* A **19**, 537–56.

— 1987b. The Church's many mansions: the changing structure of the Church Commissioners land and property holdings, 1948–1977. *Institute of British Geographers, Transactions* **12**, 465–81.

Hamilton, P. 1990. Sociology: commentary and introduction. See Lowe & Bodiguel (1990), 225–31.

Harvey, D. 1993. The UK in world and European agriculture. Proceedings from the Great North Meet: the Agricultural Conference for North Britain.

Hawkins, E., J. Bryden, N. Gilliatt, N. MacKinnon 1993. Engagement in agriculture 1987–1991: a West European perspective. *Journal of Rural Studies* **9**(3), 277–90.

Healey, P., J. Davis, M. Wood, M. Elson 1982. *The implementation of development plans: report of an exploratory study for the DoE*. Department of Town Planning, Oxford Polytechnic.

— J. Davis, M. Wood, M. Elson 1985. *The implementation of planning policies and the role of development plans, Vol 1: main findings*. Oxford, Department of Town Planning, Oxford Polytechnic.

— P. Macnamara, M. Elson, J. Doak 1988. *Land use planning and the mediation of urban change*. Cambridge: Cambridge University Press.

Herrington, J. 1984. *The outer city*. London: Harper & Row.

Hepworth, M., A. Green, A. Gillespie 1987. The spatial division of information labour in Great Britain. *Environment and Planning* A **19**, 793–806.

Hindess, B. 1987. *Politics and class analysis*. Oxford: Basil Blackwell.

Hirsch, F. 1978. *The limits to growth*. Cambridge, Mass.: Harvard University Press.

Hodge, I. 1990. Land use by design. See Britton (1990), 105–18.

House Builders Federation 1986. *Economic growth and planning in the South East*. London: HBF.

Howells, J. 1984. The location of research and development: some observations and evidence from Britain. *Regional Studies* **18**(1), 13–29.

IER 1988. *Institute for Employment Research projections*. University of Warwick.

Bibliography

Jones Lang Wootten 1987. *The decentralization of offices from Central London.* London: Jones Lang Wootten.

Kneale, J., P. Lowe, T. Marsden 1991. *The conversion of agricultural buildings: an analysis of variable pressures and regulations towards the post-productivist countryside.* Working Paper 29, ESRC Countryside Change Initiative, Department of Agricultural Economics, University of Newcastle upon Tyne.

Lewis, E. 1962. A geographical description of mid-Buckinghamshire and Aylesbury. Unpublished paper, Aylesbury County Library.

Leyshon, A. & N. Thrift 1993. The restructuring of the UK financial services industry: a reversal of fortune? *Journal of Rural Studies* 9(3), 223–42.

Little, J. 1987. Rural gentrification and the influence of local-level planning. In *Rural planning: policy into action?* P. Cloke (ed.), 185–99. London: Harper & Row.

Lockwood, D. 1988. The weakest link in the chain: some comments on the Marxist theory of action. *Research in the Sociology of Work* 1, 435–81.

Lowe, P. 1977. Access and amenity: a review of local environmental pressure groups in Britain. *Environment and Planning* A 9, 35–58.

—1988. Environmental concern and rural conservation concerns. In *Land use and the European environment,* M. Whitby & S. Openshaw (eds), 68–77. London: Belhaven.

—& M. Bodiguel (eds) 1990. *Rural studies in Britain and France,* London: Belhaven.

—J. Murdoch, T. Marsden, R. Munton, A. Flynn 1993. Regulating the new rural spaces: the uneven development of land. *Journal of Rural Studies* 9(3), 205–22.

Marsden, T. 1992. Exploring a rural sociology for the Fordist transition: incorporating social relations into economic restructuring. *Sociologia Ruralis* 32(2/3), 209–30.

—& J. Little 1990. *Perspectives on the food system.* Aldershot, England: Gower.

—P. Lowe, S. Whatmore (eds) 1990. *Rural restructuring: global processes and their responses.* London: Fulton.

—J. Murdoch, P. Lowe, R. Munton, A. Flynn 1993. *Constructing the countryside.* London: UCL Press.

—K. Sullivan, V. Lingham, J. Murdoch 1991. *Planning impacts, priorities and options: report of survey for the Buckinghamshire Structure Plan Review.* Environmental Services Unit, South Bank Polytechnic.

—J. Murdoch, K. Sullivan, V. Lingham 1992. Planning for the social limits to growth. *Planner* 78(4), 6–7.

Martin, R. 1989. Regional imbalance as consequence and constraint in national economic renewal. In *The restructuring of the UK economy,* F. Green (ed.), 80–97. London: Harvester Wheatsheaf.

Mason, C. 1985. The geography of "successful" small firms in the UK. *Environment and Planning* A 17, 1499–1513.

Massey, D. 1984. *Spatial divisions of labour.* London: Macmillan.

Massey, D. 1991. A global sense of place. *Marxism Today* (June), 24–9.

Mertz, S. 1992. The European Economic Community Directive on Environmental Assessments: how will it affect United Kingdom developers? *Journal of Planning Law,* 483–98.

Mormont, M. 1990. Who is rural? Or, how to be rural. See Marsden et al. (1990), 21–44.

Morrey, M. 1990. Long live landfill! *Roots* 7, 5–7.

Mullins, M. 1991. The identification of social forces in development as a general problem in sociology: a comment on Pahl's remarks on class and consumption relations as forces in urban and regional development. *International Journal of Urban and Regional Research* 15(1), 119–26.

Murdoch, J., J. Kneale, P. Lowe, T. Marsden 1992. *Making a market: the case of the 1980s golf course boom*. ESRC Countryside Change Working Paper 34, Department of Agricultural Economics, University of Newcastle upon Tyne.

Murdoch, J. & T. Marsden 1992. A fair way to plan? Assessing the golf course boom. *Planner* 78(16), 8–10.

Murray, R. 1988. "Crowding out": boom and crisis in the South East. Text of the first SEEDS lecture delivered at the AGM of SEEDS, London, 28 November.

Nationwide Building Society 1989. *Local area housing statistics*. London: Nationwide Building Society.

Newby, H. 1980a. Urbanization and the rural class structure: reflections on a case study. In *The rural sociology of the advanced societies*, F. Buttel & H. Newby (eds), 255–78. London: Croom Helm.

Newby, H. 1980b. *Green and pleasant land? Social change in rural England*. London: Hutchinson.

Newby, H., C. Bell, D. Rose, P. Saunders 1978. *Property, paternalism and power*. London: Hutchinson.

North, J. 1990. Future agricultural land use patterns. See Britton (1990), 69–93.

Pahl, R. 1965. Class and community in English commuter villages. *Sociologia Ruralis* 5, 5–23.

—1970. *Readings in urban sociology*. Oxford: Pergamon.

—1989. Is the emporer naked? Some questions on the adequacy of sociological theory in urban and regional research. *International Journal of Urban and Regional Research* 13(4), 711–20.

—1993. Does class analysis without class theory have a promising future? A reply to Goldthorpe and Marshall. *Sociology* 27(2), 253–58.

Parry, G., G. Moyser, N. Day 1992. *Political participation and democracy in Britain*. Cambridge: Cambridge University Press.

Peck, J. & A. Tickell 1992. Local modes of social regulation? Regulation theory, Thatcherism and uneven development. *Geoforum* 23(3), 347–64.

Phillips, M. 1993. Rural gentrification and the process of class colonization. *Journal of Rural Studies* 9(2), 123–40.

Przeworski, A. 1977. Proletariat into class: the process of class formation from Karl Kautsky's "The class struggles" to recent controversies. *Politics and Society* 7, 343–401.

Raban, J. 1974. *Soft city* London: Hamish Hamilton.

Reade, E. 1987. *The British town and country planning system*. Milton Keynes: Open University Press.

Regional trends (various). London: HMSO.

Royal and Ancient Golf Club of St Andrews 1989. *The demand for golf*. St Andrews: Royal & Ancient.

Saunders, P. 1990. *A nation of homeowners*. London: Routledge.

—H. Newby, C. Bell, D. Rose 1978. Rural community and rural community power. In *International perspectives in rural sociology*, H. Newby (ed.), 63–78. Chichester: John Wiley.

Savage, M., J. Barlow, P. Dickens, T. Fielding 1992. *Property, bureaucracy and culture: middle-class formation in contemporary Britain*. London: Routledge.

—& A. Warde 1993. *Urban sociology, capitalism and modernity*. London: Macmillan.

—P. Dickens, T. Fielding 1988. Some social and political implications of the contemporary fragmentation of the "service class" in Britain. *International Journal of Urban and Regional Research* 12(3), 455–76.

Bibliography

Savage, R. 1987. *The future of rural planning and the environmental implications: the rural view*. Royal Town Planning Institute Council Paper. London: RTPI.

Schaffer, D. 1991. Post-suburban America. *Built Environment* **17**(3/4), 189–231.

SEEDS 1987. *The South–South divide*. London: SEEDS Project.

SERPLAN 1985. *Regional trends in the South East: South East regional monitor*. London: SERPLAN.

— 1986. *Regional trends in the South East: South East regional monitor*. London: SERPLAN.

— 1987. *Guidelines for waste disposal planning in the South East*. London: SERPLAN.

— 1988. *Regional trends in the South East: South East regional monitor*. London: SERPLAN.

— 1989. *Regional trends in the South East: South East regional monitor*. London: SERPLAN.

— 1990. *A new strategy for the South East*. London: SERPLAN.

— 1992. *Regional trends in the South East: South East regional monitor*. London: SERPLAN.

Short, J., S. Fleming, S. Witt 1986. *Housebuilding, planning and community action*. London: Routledge & Kegan Paul.

Shucksmith, M. 1990. *Housebuilding in Britain's countryside*. London: Routledge.

— & M. Winter 1990. The politics of pluriactivity in Britain. *Journal of Rural Studies* **6**(4), 429–36.

Simmie, J. 1971. Public participation: a case study of Oxfordshire. *Royal Town Planning Institute, Journal* **57**, 161–2.

— 1974. *Citizens in conflict: the sociology of town planning*. London: Hutchinson.

Sports Council 1989. *Providing for golf in the southern region*. Reading: Sports Council.

Sports Council South West 1990. *A strategy for the provision of golf courses*. Crewkerne, Wiltshire: Sports Council South West.

Town and Country Planning Association 1985. *The Western Corridor*. London: TCPA.

Thomas, D. 1990. The edge of the city. *Institute of British Geographers, Transactions* **15**, 131–8.

Thompson, E. P. 1968. *The making of the English working class*. Harmondsworth: Penguin.

Thrift, N. 1987a. An introduction to the geography of late twentieth century class formation. In *Class and space*, N. Thrift & P. Williams (eds), 207–53. London: Routledge.

— 1987b. Manufacturing rural geography. *Journal of Rural Studies* **3**(1), 77–81.

— 1989. Images of social change. In *The changing social structure*. C. Hamnett, L. McDowell, P. Sarre (eds), 13–42. London: Sage.

Touraine, A. 1988. *Return of the actor*. Minneapolis: University of Minnesota Press.

Van der Ploeg, J. D. 1990. *Labour, markets and agricultural production*. Boulder, Colorado: Westview.

Warnes, A. 1991. London's population trends: metropolitan area or megalopolis? In *London: a new metropolitan geography*, K. Hoggart & D. Green (eds), 159–71. London: Edward Arnold.

Who's who in industry 1991. London: Fulcrum.

Williams, V. 1993. The village people. *Guardian* (13 November).

Index

Printed and bound by CPI Group (UK) Ltd, Croydon, CR0 4YY

23/10/2024

01777665-0005